PRIDE

XANE VOLUME 10

CAN XANE TAME PRIDE BEFORE IT CAUSES FAILURE? ?

Copyright

Dedication

Dedicated to my beloved grandmother Leila, whose unwavering faith nurtured my own spiritual growth. I will forever remember the praying hands that graced her television, a symbol of her enduring strength and guidance.

Contents

1. Xane 1
2. Leila 6
3. Xane 10
4. Ruby 15
5. Xane 20
6. Ruby 26
7. Lena, ... 31
8. Xane 35
9. Ruby 40
10. Xane 45
11. Leila 50
12. Ruby 55
13. Lena 59
14. Xane 64
15. Xane 68
16. Zoe 73
17. Leila 77
18. Lena 83
19. Xane 86
20. Xane 92
21. Xane 97
22. Xane 101
23. Zoe 107
24. Xane 112
25. Ruby 116
26. Xane 120

27. Ruby 124
28. Zoe... 129
29. Leila.. 133
30. Zoe... 137
31. Xane 141
32. Ruby 144
33. Xane 148
1. Xane 1
2. Xane 8
3. Xane 18
4. Xane 25
5. Xane 30

1. Xane

Sitting on the edge of the stage, I watched the others slowly trickle out of the rehearsal room, their voices quieter than usual. The weight of what had just happened still lingered in the air like a bad chord. Zoe's words, sharp and filled with a pride that bordered on arrogance, had cut deeper than I think she realized. Lena's hurt had been visible, even as she tried to keep it together.

Zoe's words *echoed* in my mind after her criticism of Lena. She had scoffed and rolled her eyes, saying, "Oh, please. The band's sound doesn't hinge on you keeping up. We could easily find someone else to play those keyboard parts."

I shifted slightly, feeling a familiar presence settle into my mind. The Analyst's voice was calm, almost like she'd been waiting for this moment.

"Xane," she began, *"your next emotional challenge will center around understanding pride."*

I sighed softly, resting my elbows on my knees as I stared out at the empty room. Pride. It was an emotion I had noticed creeping into my interactions with the band recently—small moments of satisfaction when we nailed a performance, and larger moments like today when Zoe's pride turned into something much darker.

"Pride can be both empowering and destructive," the Analyst continued. *"In its healthy form, it pushes people to excel and value their accomplishments. But in its unhealthy form, it leads to arrogance, division, and isolation."*

I nodded, knowing she was right. Today had been a clear example of that. Zoe was one of the most talented drummers I'd ever played with, and her dedication to the band was undeniable. But somewhere along the way, her pride had shifted from feeling accomplished to needing control—needing to prove she was indispensable, even at the expense of the rest of us.

"How do I start?" I asked quietly, more to myself than to the Analyst. I'd learned so much already—*fear, shame, sadness, love, anger, and even envy*—but this felt different. Pride was more subtle, more complex. It could build someone up, but it could also destroy relationships if left unchecked.

"You begin by observing it in yourself and others," the Analyst said. *"Notice when pride motivates someone to strive for excellence, and when it crosses the line into hubris."*

I thought back to Zoe, remembering the look in her eyes during rehearsal—defensive, almost cold. She hadn't always been like that. I'd seen the joy in her when we first started playing together, the pride she took in her work when it wasn't tinged with arrogance. But today, something had shifted. She wasn't just proud of her abilities—she needed to assert them, to prove something to herself and everyone else.

"I think Zoe's stuck in that negative space," I murmured. "It's like she's using pride to push everyone else away."

"That's common," the Analyst replied. *"Negative pride—hubris—often isolates people. It creates walls between them and others, making them feel like they must always be the best, always be in control. But remember, pride isn't inherently bad. It's about balance."*

I leaned back against the stage, letting the words sink in. Balance. It always came back to that, didn't it? Every emotion I'd learned so far had been about balance—understanding the good and the bad, and finding a way to navigate between them. Pride was no different. In some ways, it could push us to be better, to feel good about our accomplishments, but if it went unchecked, it could easily tip into something destructive.

As I thought about my own experiences, I realized that I had felt moments of pride, too—especially recently. I was proud of how far the band had come, proud of the music we were making, proud of the way we'd grown together. But I had to be careful. I didn't want my pride to turn into the kind of arrogance Zoe had displayed today.

"What's the difference between healthy pride and hubris?" I asked aloud, knowing the Analyst would answer.

"Healthy pride is grounded in humility," she explained. *"It allows you to take pride in your accomplishments while acknowledging the efforts of others. It builds connections rather than tearing them apart. Hubris, on the other hand, is self-centered. It's the belief that you are better than others, that you don't need anyone else. It isolates, creating an illusion of superiority."*

I closed my eyes, thinking about how pride had played out in my own life. I was proud of what we'd achieved as a band, but I'd always seen it as our success, not just mine. Was that the key? Seeing pride as something shared rather than something claimed for yourself?

"I think I'm starting to understand," I said softly, more to myself than to the Analyst.

"Good," she replied. *"This journey will require you to not only understand pride in yourself but also in those around you. Zoe, for example, is struggling with hubris. But perhaps she doesn't realize it yet."*

I nodded. Zoe's actions today had been hurtful, but I knew that underneath it, she wasn't a bad person. She was just caught in a cycle of needing to prove herself, and that was a dangerous place to be.

As the Analyst's voice faded into the background, I stood up, feeling a strange mix of excitement and caution. This was going to be a challenging mission—pride wasn't as obvious as anger or sadness. It was subtle, often masked as something else. But I was ready to learn.

I glanced toward the rehearsal room door, where Zoe had disappeared earlier. This journey was just beginning, and it wasn't going to be easy.

But nothing worth learning ever was.

2. Leila

I could barely keep the grin off my face as I walked into the rehearsal room. The news I was about to share was going to change everything. We'd had some amazing gigs recently, but this—this was on a whole new level.

"Alright, everyone, listen up!" I called out, clapping my hands to get the band's attention.

Ruby glanced up from tuning her bass, a curious smile playing on her lips. Zoe was already seated at her drum kit, her sticks in hand, tapping out a quick rhythm, while Lena fiddled with her keyboard, clearly lost in her own world. Xane, of course, was standing off to the side, quiet but attentive as always.

When everyone had settled down, I took a deep breath, trying to contain my excitement. "You're not going to believe this, but… we've just been chosen to have one of our songs lead in for a major movie!"

The room exploded into a mix of shouts and cheers. Ruby let out a loud whoop, jumping up to high-five me, while Lena's face lit up in disbelief. Xane smiled, her usual calm shifting into something brighter, and even Zoe's eyes widened with excitement.

"A movie?" Ruby grinned. "Like, a real movie? No way!"

"Yes, way!" I laughed. "It's a huge opportunity, guys. Our song is going to be the lead-in track. Millions of people are going to hear it."

Lena shook her head, still processing the news. "I can't believe this is happening."

"We made it," Zoe said, her voice quieter but carrying a note of pride. "We're really doing it."

I could see the gears turning in everyone's heads. This was the kind of break every band dreamed of—the chance to reach an audience far beyond anything we'd ever imagined. It was going to catapult us into a whole new level of success.

"Alright, let's not get ahead of ourselves," I said, though I couldn't stop smiling. "We've got to stay focused. There's a lot to do. We'll be working with a film crew, and they want to shoot a music video for the track."

Ruby practically bounced with excitement. "A music video? This just keeps getting better!"

But as the initial excitement started to settle, I noticed Zoe's expression shift. Her pride was still there, but something darker was starting to creep in—a kind of tension that hadn't been there before. Her hands tightened around her drumsticks, her posture a little more rigid.

I knew what she was thinking. The stakes were higher now. This wasn't just about making music anymore—it was about proving we belonged here. And for Zoe, that pressure could turn pride into something harder to control.

"Zoe, you alright?" I asked, trying to keep my tone light.

She glanced at me, her smile a little forced. "Yeah. I just… we can't afford to mess this up."

"No one's going to mess this up," I reassured her. "We've got this. We're a team, remember?"

She nodded, but the tension in her shoulders didn't ease. I made a mental note to keep an eye on her. Zoe was proud of the band—of what we'd achieved—but I could tell she was already putting too much pressure on herself. That kind of pride could easily turn into something dangerous if she wasn't careful.

As we started rehearsing, I tried to push the thought aside and focus on the music. This was the chance of a lifetime, and I wasn't going to let anything ruin it. We'd made it this far together, and I knew we could handle whatever came next.

But as I caught a glimpse of Zoe's clenched jaw and the intensity in her eyes, I couldn't shake the feeling that this was just the beginning of a new challenge—one that might test more than just our music.

3. Xane

The moment we stepped onto the film set, it felt like we'd entered another world. The air was buzzing with activity—cameras being adjusted, lights flickering on and off, and people shouting instructions from all directions. It was pure, unfiltered chaos.

Leila looked over at me, raising an eyebrow. "You ready for this?" she asked with a grin.

I shrugged, offering a small smile in return. "I've learned to expect the unexpected."

Ruby, always ready for excitement, bounced on her toes as she took in the scene. "This is going to be wild," she said, practically vibrating with energy. "We're shooting a real music video for a movie! How cool is that?"

Lena, on the other hand, looked more hesitant, glancing around with wide eyes. Zoe, though… Zoe was different. There was a confidence in her that had always been there, but today, something about it felt off. It wasn't just confidence—it was bordering on arrogance.

We were introduced to the director, a guy named Victor, who didn't even bother looking up from his clipboard as he gave us a rushed greeting. "Alright, we've got a lot to do and not much time," he said, waving us toward the set. "Let's get this going. I need everything perfect."

Perfect. The word hung in the air like a warning.

We followed the crew to the set, where actors dressed in elaborate costumes were rehearsing their scenes. The actors, larger-than-life personalities, barely acknowledged us as they passed by, talking loudly about their next big projects. I caught Ruby mouthing "divas" to Leila, who stifled a laugh. Even I had to admit, the atmosphere felt stifling.

"Okay, band," Victor barked, clapping his hands. "We're going to shoot the performance scene first. We need energy. Big energy. Got it?"

Zoe stepped forward, her drumsticks in hand, her expression already hardening. "Don't worry, we've got this," she said, her voice a little too sharp. "We're professionals."

Victor barely glanced at her. "Great. Just make sure you hit your marks."

As we set up our instruments, I could feel the tension growing. The director was demanding, the actors were dismissive, and Zoe—Zoe was starting to show the signs of cracking. Her confidence, which had always been one of her strengths, was quickly shifting into something more aggressive.

We began playing through the song, trying to find a rhythm, but every time we'd get into it, Victor would cut us off. "No, no, no!" he shouted. "More energy! You need to be bigger!"

Zoe's jaw clenched tighter with every interruption, and I could see her frustration mounting. She slammed her drumsticks down on the set, making a loud crack that echoed through the room. "We're giving you energy," she snapped. "Maybe if you gave clearer instructions, we wouldn't have to keep stopping."

Victor shot her a look. "Listen, I'm running this shoot. You play the music, I'll direct the video. Got it?"

The tension between them was electric, and for a second, I thought Zoe was going to fire back. She opened her mouth, but Leila, ever the peacemaker, stepped in.

"Let's all take a breath," Leila said, her voice calm. "We're here to make something great, right? Let's work together."

Zoe didn't say anything, but I could see the tightness in her posture. This wasn't the Zoe I knew—the one who loved drumming for the pure joy of it. This was someone who felt like she needed to prove herself, to stand out, even if it meant clashing with everyone around her.

As the day dragged on, things didn't get any easier. The actors, clearly feeling the pressure of the shoot, kept flubbing their lines, leading to more delays. The wardrobe department couldn't seem to get our outfits right, and we were forced into multiple costume changes. Each time we had to stop and reset, I could see Zoe's patience wearing thin.

By the time we wrapped for the day, she was practically fuming. "This is ridiculous," she muttered as we packed up our gear. "We're a band, not some backup act for a bunch of prima donnas."

"Hey, it's just one video," Ruby said, trying to lighten the mood. "Besides, you're totally rocking those costumes. You look like a drumming superhero."

Zoe didn't even crack a smile. "I don't care about the costumes. We're better than this. I'm better than this."

I exchanged a glance with Leila. That was the moment I realized what was really happening. Zoe's pride had shifted. It wasn't just about being good anymore—it was about being better than everyone else, about proving she was the best. And that was a dangerous road to go down.

The Analyst's voice echoed in my mind, calm and steady as always. "Negative pride isolates people, Xane. Zoe is losing sight of what's important—the team, the music. She's letting her pride take control."

I sighed inwardly, knowing the Analyst was right. This shoot was supposed to be an exciting opportunity, a chance for us to show the world what we could do. But if Zoe didn't find a way to rein in her pride, it could tear us apart before we even made it to the big screen.

Tomorrow would be another day of shooting, and I could only hope things would get better.

But deep down, I knew the real challenge had only just begun.

4. Ruby

I had to admit, I was excited about the shoot. I mean, how often do you get to be part of a music video for a major movie? It was supposed to be fun, a big step forward for the band, but instead, it felt like we'd walked onto the set of a never-ending drama—starring a cast of divas.

I was doing my best to keep things light, cracking jokes whenever the tension started to build, but it was hard when everyone around me was acting like they were auditioning for an award for "Most Difficult to Work With."

"Xane, I'm dying in this outfit," I whispered to Xane as I adjusted the ridiculous costume they'd shoved me into. It was supposed to look rock 'n' roll, but it felt more like I was dressed for a space opera. "Do you think they've got AC on this set? Because I'm pretty sure I'm going to melt."

Xane chuckled softly, shaking her head. "You'll survive."

"Barely," I muttered, glancing toward the far end of the set where a group of actors were gathered, their voices rising in a mix of complaints and demands. One in particular, a guy named Tyler who was playing the lead in the movie, was throwing a fit because his hair wasn't "perfect" for the scene.

"Can someone get me a new stylist?" Tyler snapped, waving his arms in the air dramatically. "This is unacceptable. I can’t work like this."

I rolled my eyes, trying not to laugh out loud. "Unbelievable," I whispered to Leila, who stood next to me. "This guy’s losing it over a hairbrush."

Leila snorted, her eyes twinkling with amusement. "You think we’ll get out of here before the next decade?"

Just then, Tyler stormed over to where we were standing, clearly in the middle of a meltdown. "Can you believe this?" he huffed, glaring at his reflection in a nearby mirror. "I’ve never worked under such conditions. My entire look is ruined."

I raised an eyebrow, giving him my best sympathetic nod. "Yeah, that sounds… really rough."

Tyler squinted at me, clearly not picking up on my sarcasm. "You don’t get it," he said, crossing his arms. "When you’re the star, everything matters. The details make or break you."

"I guess," I said, biting my lip to keep from laughing. "Good thing you’ve got us here to back you up."

Tyler finally seemed to realize he was talking to the band, and his expression shifted, a smirk spreading across his face. "Oh, you guys are the band, right?

Cool. Just make sure not to mess up my entrance when you play. Timing is everything in showbiz."

"Don't worry," I replied with a grin. "We'll make sure to match your hair-flipping choreography perfectly."

Tyler gave me a confused look, clearly not catching on to the joke. He walked off, still muttering about his hair, and I turned to the others with a dramatic sigh. "Ladies and gentlemen, I give you: the Diva of the Day."

Leila chuckled, shaking her head. "He's something else."

"Too bad the Academy doesn't give out awards for 'Most Dramatic Actor in a Music Video,'" I added, making Xane laugh.

But as I was soaking up the amusement of the situation, my eyes drifted over to Zoe. She was seated at her drum kit, but there was something different about her posture—something more rigid, more intense. I couldn't put my finger on it right away, but it felt… off.

"Zoe, you good?" I asked, walking over to her.

She didn't look up, just kept tapping her drumsticks on the edge of her snare, the sound sharp and repetitive. "I'm fine. Just getting ready."

I glanced at Leila, who shot me a look that said she'd noticed the same thing I had. Zoe's tone was clipped, her eyes focused but a little too intense. It was like she was gearing up for a competition instead of a music video.

"Take it easy," I said, trying to keep things light. "It's just a video, not a battle of the bands."

Zoe finally looked up, her expression hard. "We need to get this right. If we mess up, it's on us, not them. And I'm not here to be second-best."

I blinked, taken aback by the harshness in her voice. This wasn't the Zoe I knew—the one who quietly held the band together with her steady beats and calm demeanor. No, this was someone else entirely—someone with something to prove.

"Whoa, slow down," I said, raising my hands. "We're all on the same team here. It's just a music video, not life or death."

Zoe's jaw clenched. "That's easy for you to say, Ruby. You're not the one keeping the rhythm. If I'm off, the whole thing falls apart."

Leila, sensing the tension, stepped in. "Zoe, relax. You're the best drummer out there. We know that. Just don't put too much pressure on yourself."

But Zoe didn't seem to hear her. She was too focused, too wound up. And I couldn't help but notice how her frustration with the situation mirrored the very divas we'd been joking about earlier. It was like Zoe was becoming what she despised—someone who let pride push her too far.

I exchanged a quick glance with Xane, who was watching Zoe carefully, her eyes thoughtful. I could tell she was processing something—probably about how Zoe's confidence was morphing into something dangerous.

"Zoe," I said softly, "we've got this. You don't have to carry it all by yourself."

For a split second, I thought I saw something flicker in Zoe's expression—something softer, more vulnerable—but then it was gone, replaced by that same rigid determination.

"Yeah," she muttered, more to herself than to me. "I know."

But I wasn't sure she really did.

As we geared up for the next take, I couldn't shake the feeling that Zoe's pride was leading her down a path that none of us could follow. And if she didn't figure it out soon, the video shoot might be the least of our worries.

5. Xane

The tension on set had been building all day. Between the actors throwing tantrums over costumes, the director barking orders like we were soldiers instead of musicians, and the constant delays, it felt like we were all walking on a tightrope. And Zoe… Zoe was about to lose her balance.

I could see it in the way her jaw clenched every time Victor, the director, stopped us mid-song to nitpick something small. Her drumming, usually so fluid and rhythmic, had become stiff—like she was trying to control every beat, every moment. The more she tried to force it, the more things seemed to fall apart.

"Alright, let's go again," Victor called out, looking completely unfazed by the frustration spreading through the room. "Zoe, can you bring the tempo up this time? It's dragging."

Zoe's eyes narrowed as she gripped her drumsticks tighter. "I'm not dragging. Maybe you're just not hearing it right," she shot back, her voice cold.

Victor raised an eyebrow, clearly not used to being challenged. "I am hearing it right. And you need to listen."

I winced, knowing this wasn't going to go well. Zoe wasn't someone who liked being told she was wrong, especially not in front of the whole band, crew, and cast.

She didn't respond, just turned back to her drums, but I could see her fingers trembling ever so slightly as she adjusted her grip on the sticks. Ruby, always trying to lighten the mood, gave her a playful nudge. "Come on, Zoe, it's just one more take. We'll nail it."

Zoe didn't even acknowledge her. Her focus was laser-sharp on the drums, her back straight, her shoulders tense. The moment Victor called for action, she launched into the beat with a fierceness I hadn't seen before. It wasn't just energy—it was frustration, anger, and pride all rolled into one.

As we played, I could feel the shift. Her drumming was technically perfect, but there was something off about it—something that felt too rigid, too forced. And I wasn't the only one who noticed. Victor stopped us halfway through, throwing his hands up in exasperation.

"Zoe, you're playing like a robot. We need feeling, not aggression."

The room went still, everyone waiting for Zoe's reaction. I could see it building in her—an explosion waiting to happen. She stood up, throwing her drumsticks down with a loud clack that echoed across the set.

"You don't know what you're talking about," she snapped, her voice dripping with anger. "I'm playing it exactly how it should be played. Maybe if you'd give clearer direction, we wouldn't have to keep stopping!"

Victor crossed his arms, his expression calm but icy. "I've been doing this for years. I know what I'm talking about. You're not giving me what I need."

Zoe's face flushed red as she took a step forward, her body tense. "Maybe what you need is to stop micromanaging every second of this shoot and let us do what we're good at."

I held my breath, glancing at Leila and Ruby, who looked just as shocked as I felt. Zoe never lost her temper like this—not with us, not with anyone. But now, it was like she was barely holding herself together.

The Analyst's voice echoed softly in my mind, a calm presence in the middle of the chaos. *"Pride is driving her, Xane. She's letting it take over—turning confidence into arrogance."*

I knew she was right. Zoe's pride wasn't just about her drumming anymore—it had become something else, something darker. She was letting her need to prove herself consume her, and it was starting to tear her apart from the inside.

"Zoe," I said quietly, stepping forward, "let's take a break. We can figure this out."

But Zoe didn't even look at me. She was too far gone, her anger blazing as she squared off with Victor. "You don't know what you're talking about," she repeated, her voice louder, more defiant.

Victor, clearly done with the back-and-forth, waved a hand dismissively. "Fine. Take five. We'll start again when you're ready to act like a professional."

Zoe let out a frustrated sound, grabbing her drumsticks and storming off the set without another word. The rest of us stood there, awkwardly silent as the crew exchanged nervous glances. Ruby gave me a worried look, and I could see the question in her eyes—what's going on with Zoe?

I didn't have an answer. All I knew was that Zoe's pride had shifted from something healthy and productive into something destructive. She wasn't just proud of her talent—she was letting it control her, making her feel like she had to fight everyone to prove her worth.

I followed Zoe outside, where she was pacing, her expression dark and brooding. I didn't say anything at first, just stood there, watching as she wrestled with whatever was going on inside her.

After a moment, she stopped and looked at me, her face hard. "What?"

I met her gaze evenly. "You need to calm down."

She scoffed, shaking her head. "Calm down? I'm fine."

"No, you're not," I said, keeping my voice steady. "You're letting your pride get in the way, Zoe. You're trying to prove something, and it's tearing you apart."

Her eyes flashed with anger. "I don't need you to analyze me, Xane."

"I'm not analyzing," I replied, stepping closer. "I'm telling you what I see. And right now, I see someone who's letting their pride control them."

For a second, I thought she was going to lash out again, but instead, she looked away, her shoulders slumping slightly. "I just… I can't mess this up. I can't let anyone think I'm not good enough."

I softened, understanding washing over me. "You don't have to prove anything, Zoe. We know you're good enough. You just need to stop letting pride make you think otherwise."

She didn't respond right away, but I could see the gears turning in her mind, the tension slowly easing from her body. The Analyst's voice whispered in the back of my mind, reminding me that this was only the beginning of Zoe’s struggle with pride—and it was going to be a long road ahead.

But for now, I was hopeful that she could find her way back.

"You ready to go back in?" I asked gently.

Zoe sighed, nodding. "Yeah. Let’s just get this over with."

As we walked back onto the set, I couldn’t shake the feeling that this was far from over. Pride, in all its forms, was something we’d all have to face.

And Zoe’s battle with it was just beginning.

6. Ruby

Everything seemed to be going okay, for the most part. I mean, considering we were filming a high-profile music video in the middle of chaos central, I'd say we were holding it together. The earlier tension had simmered down, and Zoe, while still a bit moody, was back at her drum kit, ready for the next take.

But then it happened.

We were mid-song, the cameras rolling, and for the first time all day, it actually felt like we were in the zone. The music was tight, the energy was up, and I could see the crew nodding along, finally satisfied with our performance.

Then, out of nowhere, I heard a rip.

At first, I thought it was just part of the sound effects or something in the background. But the look on Zoe's face told me everything I needed to know. Her eyes went wide, and for a split second, she stopped playing, her gaze darting down to her outfit.

I followed her eyes and—yep. There it was. A huge tear down the side of her shirt, exposing the top half of her drumstick tattoo, and definitely not in the cool, rockstar way we were going for.

"Oh no," I muttered under my breath, biting my lip to keep from laughing. This was a classic wardrobe malfunction if I'd ever seen one.

"Cut!" Victor yelled, throwing his hands in the air like the world had ended. "What was that? What's going on?"

Zoe immediately stood up, her face flushed with embarrassment. "Are you kidding me? This is ridiculous!" she shouted, holding the torn fabric together with one hand. "What kind of wardrobe department lets this happen?"

I could see the panic and anger rising in her, and I knew this wasn't just about the shirt. Zoe had been walking a fine line between confidence and arrogance all day, and now that something had gone wrong—something completely out of her control—she was losing it.

"Zoe, relax," Leila said calmly, stepping in. "We'll get it fixed. It's not a big deal."

"Not a big deal?" Zoe snapped, her voice sharp enough to cut glass. "We're filming a video that's going to be seen by millions of people! I can't have this happen!"

"Come on, it's just a shirt," I said, trying to lighten the mood. "Besides, it's kind of punk rock, don't you think?"

But Zoe wasn't having it. She glared at me, her pride clearly wounded. "I can't believe this. This is a complete disaster."

I glanced at Xane, who was watching Zoe with that intense, thoughtful look she always got when she was reading into something deeper. I knew what she was thinking—that this wasn't about the shirt at all. Zoe's pride had been slowly taking over all day, and now that something had gone wrong, she couldn't handle it.

This wasn't the first time I'd seen Zoe react like this, but it was different now. Before, she'd always been able to brush things off, roll with the punches. But lately, every little mistake seemed to dig at her, like she had to be perfect, had to prove herself at every moment.

"Okay, everyone, take five," Victor said, waving the crew away as he muttered something about wardrobe emergencies. The rest of the band exchanged glances, unsure of how to diffuse the situation without making it worse.

"Zoe," Leila said softly, "it's just a shirt. You're still you. We're still the band. This doesn't change anything."

But I could see Zoe wasn't hearing her. Her pride had already clouded her judgment, making this minor malfunction feel like a massive failure. She turned on her heel, marching off to the wardrobe area with her head held high, but I could see the tightness in her posture, the frustration rolling off of her.

"Wow," I said to Xane, who had come up next to me. "I knew Zoe could be intense, but this is… something else."

Xane nodded, her eyes following Zoe's retreating figure. "She's letting her pride get in the way."

I sighed, shaking my head. "Yeah, but it's just a shirt. We've all had wardrobe malfunctions before."

"It's not about the shirt," Xane said quietly. "It's about control. Zoe's trying so hard to stay in control that the smallest thing going wrong feels like everything's falling apart."

I looked back toward the wardrobe area, where Zoe was undoubtedly chewing out the poor costume team for the rip in her shirt. It was clear that today wasn't just about making a great video for Zoe—it was about proving she could do it all perfectly.

And that kind of pride? It was dangerous.

"We've got to help her out," I said, running a hand through my hair. "She's going to burn out if she keeps this up."

Xane nodded again, a thoughtful expression on her face. "She's not ready to listen yet. But we'll be here when she is."

I sighed again, glancing at the others, who were watching the scene unfold with quiet concern. This video shoot was supposed to be the start of something huge for us—a stepping stone to bigger and better things. But if Zoe didn't get a handle on her pride soon, it could tear us apart before we even made it to the premiere.

"Here's hoping she figures it out soon," I muttered, grabbing a bottle of water from the table and taking a long drink. "Because this whole 'diva' thing isn't a good look on her."

Xane gave me a small smile, but I could see the worry in her eyes. We both knew Zoe wasn't just fighting a wardrobe malfunction. She was fighting herself.

And that was a battle only she could win.

7. Lena,

I watched from the corner of the set as Zoe stormed off toward the wardrobe area, her face flushed with frustration. She'd been on edge all day, snapping at everyone and acting like the weight of the world rested on her shoulders. It wasn't just today, though. Zoe had been like this for weeks—ever since we got the news about the music video.

I pressed the keys on my keyboard softly, trying to distract myself from the knot of anxiety tightening in my stomach. The feeling of being... small. Pushed aside. Overlooked.

It wasn't like Zoe to act like this, but the more I thought about it, the more I realized this wasn't a one-time thing. Lately, it felt like she didn't value me, like she thought the band could move forward without me—especially after what she said during rehearsal the other day. We could get anyone to play those keyboard parts. That phrase had been ringing in my head ever since.

Was that really how she saw me? As replaceable?

I looked over at the rest of the band, Leila and Ruby talking in hushed tones while Xane stood nearby, her face thoughtful as always. They were all so... comfortable. So confident in their place here. But me? I wasn't so sure anymore.

I'd joined Sapphire Nights because music was the one thing that had always made sense to me. The melodies I played were a way of expressing things I could never put into words. But now, standing here, I couldn't help but wonder if Zoe was right. Was I holding them back?

I pressed another soft chord, listening to the way it lingered in the air. My parts had always been quieter, more subtle, but that didn't mean they weren't important. At least, I had always believed they were.

But after what Zoe said... maybe I wasn't pulling my weight like the others. Maybe I wasn't contributing as much as I thought I was.

I glanced over at Xane, who caught my eye and gave me a small, reassuring smile. Xane always had a way of making people feel seen, of making you feel like you mattered. But even her kind smile couldn't push away the doubt creeping in.

Was Zoe right?

I couldn't shake the feeling that the band didn't need me anymore. I mean, they were all so talented in their own ways. Leila's voice could fill an arena, Zoe's drumming kept us grounded, Ruby's bass line gave us that wild, carefree edge, and Xane—Xane was just a natural performer, drawing people in with her saxophone and quiet charisma.

And then there was me. The quiet one. The one in the background, blending in but never really standing out.

I loved being part of the band, but lately, it felt like I was just... there. No one really noticed when I played, and no one missed me when I didn't. Maybe Zoe was right—maybe the band would be better off without me. They could probably get some flashy new keyboardist who would bring more energy, more... something. Something I didn't have.

The thought made my chest tighten, and I had to take a deep breath to keep the tears from welling up. I didn't want to be the weak link. I didn't want to hold the band back.

But maybe that's exactly what I was doing.

I sat down on the edge of a crate, folding my hands in my lap, feeling smaller than ever. The excitement of the video shoot had been tainted by this growing sense of unease, and I didn't know how to shake it.

"Lena?" Xane's voice broke through my thoughts, and I looked up to see her standing beside me, her expression soft. "You okay?"

I nodded, forcing a smile. "Yeah, I'm fine. Just... thinking."

Xane didn't say anything for a moment, but she sat down beside me, not pushing, just... there. That was the thing about Xane—she always seemed to know when someone needed space but didn’t want to be alone.

"You know," she said quietly, "just because something’s quiet doesn’t mean it isn’t important."

I blinked, glancing at her. "What do you mean?"

She looked at my keyboard, then back at me. "Your parts—they’re subtle, but they’re the glue that holds everything together. Without you, the music would fall apart."

I swallowed, feeling the knot in my stomach loosen just a little. "I don’t know, Xane. Lately, it feels like... I’m not enough."

She met my gaze, her eyes steady and kind. "You are enough, Lena. Don’t let anyone, not even Zoe, make you doubt that."

I nodded, though the doubt was still there, lurking beneath the surface. I wanted to believe Xane, but Zoe’s words had struck a chord I couldn’t ignore.

As we sat there in silence, I couldn’t help but wonder if I’d ever find my place in the band again—or if I even had one to begin with.

8. Xane

After the long day on set, the chaotic energy still clung to me. The tension with Zoe, Lena's quiet struggle, and the stress of the video shoot all swirled in my mind. But underneath it all was something deeper—something the Analyst had been hinting at for days. Pride.

It was everywhere, simmering beneath the surface of everything we did. And after seeing Zoe's outbursts, I knew I needed to get a better handle on it before things spiraled further.

As I sat in the dim quiet of the bus, I closed my eyes, letting my thoughts settle. The Analyst's voice came through gently, like a soft breeze across my mind.

"How are you feeling, Xane?" she asked, her tone calm as ever.

I thought about the mess of emotions from the day—Zoe's frustration, Lena's doubt, and even my own feelings of pride. I was proud of how far the band had come, proud of the opportunities we were getting, but there was a part of me that worried. Was I walking the same line as Zoe? Could my own pride twist into something darker?

"I'm trying to figure it all out," I admitted. "Pride… it's tricky. It's like there's this fine line between being proud of what we've accomplished and letting that pride turn into something more—something that pushes people away."

The Analyst was silent for a moment, letting me reflect before speaking. *"You're right. Pride can be a powerful motivator, but it can also isolate. It's important to understand the difference between healthy pride and arrogance."*

I nodded, her words sinking in. "Healthy pride is feeling good about what you've achieved, right? But arrogance is when you start thinking you're better than everyone else. Like you don't need anyone."

"Exactly," the Analyst replied. *"Healthy pride is grounded in gratitude and humility. It allows you to take pride in your accomplishments without diminishing the value of others. Arrogance, on the other hand, often stems from insecurity—it's a defense mechanism used to protect a fragile ego."*

I thought about Zoe. Her confidence, once something that drove the band forward, had morphed into arrogance over the past few weeks. She wasn't just proud of her talent anymore—she was using it as a shield, as if she needed to prove she was better than everyone else. And that was pushing us all away.

"So how do I keep that balance?" I asked. "How do I stay proud of what we've done without falling into arrogance?"

"By remembering that your success is not solely your own," the Analyst explained. *"Pride becomes toxic when it separates you from others—when it makes you feel as if you've achieved everything alone. But healthy pride comes from recognizing the collective effort, from understanding that your accomplishments are part of a larger whole."*

I nodded, thinking about the band. Every one of us had contributed to our success. Leila's leadership, Ruby's wild energy, Lena's quiet strength, and even Zoe's determination—they had all played a role in getting us here. None of it would have been possible alone.

"It's about staying connected," I murmured, more to myself than the Analyst. "Keeping that sense of teamwork, of collaboration."

"Yes," she agreed. *"And it's also about being aware of when pride starts to shift. Pay attention to your thoughts, Xane. When pride feels empowering, when it fills you with gratitude and joy, it's healthy. But when it feels like you're trying to prove something—when it becomes about competition or superiority—that's when it becomes dangerous."*

I thought about the moments today when I had felt proud. When we'd finally nailed the music during the shoot, I had felt that swell of pride in what we had accomplished. But it wasn't just about me. It was about us—the band as a whole. That feeling had been good, healthy.

But then there were the other moments, like when I saw Zoe pushing Lena aside, or when I felt the pressure to make the video perfect. Those moments had twisted something in me—made me feel like I had to prove myself, like I had to be the best. That kind of pride… that was dangerous.

"I think I understand now," I said softly. "It's not about never feeling proud. It's about recognizing when that pride turns into something else."

"Precisely," the Analyst responded. *"Pride is not inherently negative. It can drive you to achieve great things, to believe in your abilities. But it must always be balanced with humility and connection to others."*

I opened my eyes, feeling a sense of clarity. Pride wasn't something to fear—it was something to manage, to understand. As long as I stayed aware of the line between healthy pride and arrogance, I could navigate this journey.

"Thanks," I said, a small smile tugging at my lips. "I think I'm ready for whatever comes next."

The Analyst's voice faded gently, leaving me with a sense of peace. This mission—learning about pride—was going to be challenging, but I wasn't facing it alone.

And as long as I kept the band, my friends, at the center of everything, I knew I'd be okay.

Tomorrow would be another day on set, and I had no doubt more challenges were waiting for us. But now, I had a better understanding of how to keep pride in check—how to let it lift me up without letting it pull me away from the people who mattered most.

9. Ruby

I could feel it coming before it even started. The tension in the air was thick enough to cut with a butter knife, and Zoe… well, she was already wound up tighter than a snare drum.

We were back on set after a short break, trying to push through the last few scenes of the music video, but everything was taking longer than it should have. The cameras weren't working right, the lighting was off, and the director, Victor, was barking orders like he was trying to lead a battlefield, not a film set.

"Okay, band! Positions!" Victor called, waving us toward our spots. "We need to redo the last shot. Zoe, I need more intensity from you on the drums. It's too flat."

I cringed as soon as the words left his mouth. Zoe had been prickly all day, and criticizing her playing was like tossing a match onto a pile of dry leaves. Sure enough, Zoe's eyes narrowed, and I could see her shoulders tense up.

"It's not flat," she shot back, her voice tight. "I'm giving it everything, and maybe if the cameras worked properly, you'd see that."

Oh boy.

I glanced at Leila, who gave me a look that screamed do something.

"Come on, Zoe," I said, stepping in with my best I'm-not-taking-this-too-seriously grin. "It's just one more take. Think of it as getting extra practice time. You'll be a drum goddess by the end of this."

I expected a chuckle, maybe a smirk—something to break the tension. But Zoe wasn't having it.

"Practice? I don't need practice, Ruby," she snapped, her voice sharper than usual. "I'm not the problem here."

My smile faltered as I took a small step back. Zoe had always been serious about her drumming, but this was different. This wasn't about the music anymore. It was about proving something—to herself, to the rest of us, maybe even to the world.

Leila, trying to keep things calm, stepped forward. "Zoe, no one's saying you're the problem. Let's just get through this. It's been a long day."

Zoe huffed, crossing her arms. "You all act like I'm overreacting, but I'm the one carrying this whole thing. Without me, we wouldn't even be here."

I exchanged a look with Xane, who was standing off to the side, quietly observing the whole scene. I could tell she was thinking about something deep, probably analyzing Zoe's every move, but right now, we didn't need analysis. We needed a lifeline.

"Hey, hey," I said, waving my hands dramatically. "We're all in this together, remember? I mean, yeah, you're amazing on the drums, Zoe, but I think the rest of us are pulling our weight too. You know, with my incredible bass-playing skills and all."

Leila rolled her eyes, and I could hear a couple of the crew members stifle laughter. I figured a little humor might take the edge off the situation, but instead, Zoe glared at me like I'd just insulted her entire existence.

"This isn't a joke, Ruby," she hissed, her voice dangerously low. "We've worked too hard to get here, and I'm not going to let a bunch of amateurs screw it up."

Ouch. Okay, so maybe that one stung a little.

Leila stepped in again, trying to be the voice of reason. "Zoe, calm down. We're all tired, but we're doing our best. Let's just finish this up so we can get out of here."

But Zoe wasn't listening. She was too far gone, her pride digging its claws into her and refusing to let go. "You don't get it," she muttered, shaking her head. "If we don't nail this, if I don't nail this, none of it matters."

I could feel the tension mount, the frustration thick in the air. Zoe was spiraling, and no matter how much I tried to make light of the situation, it wasn't working. Humor wasn't going to pull her out of this.

"Zoe, we've got this," I said softly, trying a different approach. "You're not carrying this alone. We're a team."

But she didn't hear me. Or if she did, she didn't care. Her pride had taken over, pushing her to a place where nothing and no one else mattered but her own performance.

Victor, clearly fed up with the drama, stepped in. "Alright, enough! We're doing the take. Zoe, get it together, or I'll find someone who will."

Zoe's jaw clenched so hard I thought she might break a tooth, but she didn't say anything. She just turned back to her drums, her face a mask of frustration.

I wanted to say something, to ease the tension, but I knew that nothing I said would fix this. Zoe was too deep in her own head, her pride turning everything into a battle she had to win.

As we started the take again, the air was thick with unspoken words. I played through the song, my fingers moving over the strings automatically, but my mind was elsewhere. I could feel Zoe's anger simmering just beneath the surface, threatening to boil over at any second.

We finished the take, and I expected Zoe to explode again, but instead, she sat there in silence, her hands gripping the drumsticks so tightly her knuckles were white. She didn't look at any of us. She didn't say a word. She just stared at the floor, the weight of her own pride crushing her from the inside out.

And as much as I wanted to help her, I didn't know how.

Zoe was fighting a battle none of us could see, and until she figured out how to let go of her pride, we'd all be caught in the fallout.

I glanced at Xane again, hoping she had some kind of plan. Zoe needed help—more than she realized.

And we needed to find a way to bring her back before the music, and the band, fell apart.

10. Xane

After the latest round of tension on set, I couldn't shake the feeling that something was wrong—more than just the surface-level arguments and frustrations that had bubbled up between Zoe and the rest of the band. Beneath all the chaos, there was a quieter struggle happening, one that I knew Lena was wrestling with.

I found her sitting at the edge of the set during a break, her fingers softly pressing the keys on her keyboard, but the usual confidence wasn't there. She was retreating into herself, her shoulders hunched, as if she was trying to disappear into the background.

That wasn't like Lena. She had always been quiet, yes, but her music had always spoken louder than words ever could. Now, though, it was as if even her music had fallen silent.

I walked over and sat next to her, not saying anything at first. The silence between us felt heavy, and I knew it was more than just exhaustion from the long day of filming. Lena was doubting herself. I could see it in the way she avoided eye contact, her hands resting lightly on the keys but not playing a single note.

"Lena," I said softly, keeping my voice gentle, "you've been quiet today."

She glanced at me briefly, then looked back down at her keyboard. "Yeah," she murmured. "Just... trying to stay out of the way."

I frowned, not liking the sound of that. "Out of the way? You're a part of this, Lena. You're not in anyone's way."

She shrugged, her eyes still fixed on the keys. "I don't know, Xane. Lately, it feels like... maybe I don't matter as much. Like maybe the band would be fine without me."

The words hit me hard. I hadn't realized just how deeply Zoe's earlier comment about replaceability had sunk into Lena. And as much as Lena tried to hide it, I could see the hurt there, the doubt that was eating away at her confidence.

"Lena," I said firmly, placing my hand on her arm to get her attention. "That's not true. You're just as much a part of this band as any of us. We need you."

She shook her head, her voice barely above a whisper. "But Zoe... she's right, isn't she? I mean, anyone could play these parts. I'm not exactly the loudest, or the most noticeable. I don't stand out the way you or Leila or even Ruby do."

I let out a soft sigh, understanding now just how deep Lena's feelings of insecurity had grown. "Lena, the fact that you don't 'stand out' doesn't mean you're not valuable. Every part of this band matters. Your music adds a layer, a depth that no one else could create."

She finally looked at me, her eyes filled with doubt. "But Zoe—"

"Zoe's frustrated," I interrupted gently. "And she's letting her pride get in the way of seeing what's right in front of her. But that doesn't mean what she said is true."

Lena bit her lip, her gaze dropping back to the keyboard. "I just don't know how to speak up. I'm not like the rest of you. I don't want to cause drama."

I leaned in, making sure she could see how serious I was. "Speaking up doesn't cause drama, Lena. Speaking up shows that you care about yourself and about this band. And we need your voice—whether it's through your music or your words."

She hesitated, the conflict still written all over her face. I could tell that part of her wanted to believe me, but the weight of Zoe's comment, of feeling small, was still dragging her down.

"You have to remember," I continued, "we all have something different to bring to this band. Just because you're quieter doesn't mean you don't have something valuable to say. Sometimes the quiet voices are the ones that make the biggest impact."

Lena finally met my eyes, and I could see a flicker of hope there—just a small one, but it was something. "You really think I'm important to the band?"

"I know you are," I said firmly. "You don't need to be loud to be heard, Lena. Your music, your presence—those are what make Sapphire Nights who we are."

She smiled a little, though it was still hesitant. "I'll try... I'll try to speak up more. But I don't know if Zoe will listen."

I smiled back at her. "Zoe's going through her own stuff right now, but I think if you give it time, she'll realize that she needs you, too. And no matter what happens, we're all in this together."

Lena nodded, the tension in her posture easing just a little. "Thanks, Xane. I guess I just needed to hear that."

I gave her arm a gentle squeeze before standing up. "We've all been there, Lena. And I'll be here to remind you whenever you need it."

As I walked away, I couldn't help but think about how easy it was for pride—whether it was Zoe's arrogance or Lena's insecurity—to cloud our judgment. But I knew one thing for sure: Lena wasn't replaceable. And with time, I hoped she would start to believe that, too.

For now, though, I was just glad I could help her find her voice, even if it was just a whisper. Because sometimes, the quietest voices carried the most important messages.

11. Leila

You could almost feel it coming—the moment the film director, Victor, was about to completely lose it. We had been dealing with delays all day, between the lighting issues, the sound system glitches, and a wardrobe malfunction that nearly sent Zoe over the edge. But the tension wasn't just confined to our band. The entire set felt like it was on the verge of collapse.

"Come on, people! We're behind schedule! Do you want this to be the disaster of the year?" Victor's voice echoed through the massive soundstage, his face flushed with frustration as he threw down his clipboard.

I exchanged a look with Ruby, who gave me a helpless shrug. Things were spiraling, and fast. We'd been here for hours, and nothing seemed to be going right. The crew was scrambling, the actors were fuming, and now the director was teetering on the edge of a full-blown meltdown.

"All right, everyone, let's take five," I called out, trying to inject some calm into the storm. "We'll reset and get this sorted. We've got this."

The band started to gather, all looking worn out but ready to push through. All except Zoe. She stood a few feet away, arms crossed and a look of pure annoyance on her face.

“I don’t see why we’re stopping,” Zoe snapped, shaking her head. “The delays aren’t our fault. If the crew can’t get it together, that’s on them.”

I cringed. This wasn’t the first time Zoe had let her frustration spill over into arrogance, but today, it was like she wasn’t even trying to hide it. The crew had overheard her, and I could see the irritated looks being exchanged as they struggled to reset the equipment.

“Zoe, we’re all tired,” I said, keeping my voice steady but firm. “But we’ve got to work with them, not against them. We’re all on the same team here.”

Zoe scoffed, rolling her eyes. “Yeah, well, if they were better at their jobs, we wouldn’t be stuck here for hours, would we?”

I could feel the eyes of the crew on us now, and I knew Zoe’s attitude wasn’t just alienating them—it was making things worse. Ruby stepped in, trying to lighten the mood as she always did. “Come on, Zoe. We’re rock stars, not drama queens. Let’s save the diva routine for the after-party, huh?”

But Zoe didn’t back down. “I’m not being a diva, Ruby. I just don’t see why we should have to wait around because other people can't do their jobs right."

This was bad. The crew was glaring now, and I could see the growing resentment toward Zoe. We were supposed to be professionals, and her arrogance was making all of us look bad.

I glanced over at Xane, who was watching the whole thing unfold with that calm, thoughtful expression she always had when things were getting tense. But I knew even Xane was worried about how this was going to play out.

Victor stormed over to us, his face flushed with frustration. "What's the problem now?" he barked, his eyes locked on Zoe.

"There's no problem if people can do their jobs," Zoe shot back, her voice sharp.

I could see Victor's eyes narrow, and I knew if I didn't do something fast, things were going to explode. "We're just resetting, Victor," I said quickly, stepping between him and Zoe. "Everyone's exhausted, but we're still on track."

Victor glared at Zoe for a moment longer before throwing up his hands in frustration. "Fine. Reset, but get it right. No more excuses."

He stalked off, barking orders at the crew, and I turned to face the band, letting out a breath I hadn't realized I'd been holding.

"We need to pull it together," I said, keeping my voice low but firm. "Zoe, I get that you're frustrated, but you can't take it out on the crew. We're all working toward the same goal here."

Zoe rolled her eyes again, but I could see a flicker of something in her expression—whether it was regret or just exhaustion, I wasn't sure. "Whatever. Let's just get it over with."

Ruby sighed, giving me a look that said, What can we do?

I put a hand on Zoe's shoulder, trying to soften the tension. "Look, we're all in this together. The delays suck, but we'll get through it if we work with them, not against them."

Zoe shrugged off my hand, clearly still irritated, but she didn't argue. I could tell she was barely holding it together, and I didn't blame her. This whole shoot had been a nightmare. But we couldn't afford to fall apart now, not when we were so close to finishing.

"Let's just focus on the music," I said, rallying the band. "We've got this. No more arguing, no more delays. We're going to make this happen."

The others nodded, and even Zoe looked a little less angry. I glanced at Xane, who gave me a small smile, her silent way of saying, Good job.

As we moved back into position, I hoped the next hour would go smoothly. Zoe's pride was getting in the way, and if we didn't find a way to balance it, it could tear us apart. She needed time for her emotions to chill. But for now, we just had to survive this shoot and remind ourselves why we were here in the first place.

The music. The band. That's what mattered.

And no amount of chaos on set could take that away from us.

12. Ruby

I didn't like what was happening. It felt like Zoe had gone from being our steady, calm drummer to someone I barely recognized—someone who was letting her pride get the best of her. The tension on set today had been bad, but the way she snapped at the crew, the way she talked down to all of us, was even worse. It wasn't just about the delays or the director's attitude. It was something deeper.

After the shoot wrapped up for the day, I found Xane sitting outside by the trailers, looking out at the sunset with that faraway look she got when she was thinking hard about something. I knew she'd noticed Zoe's behavior too, but Xane was the kind of person who always thought things through before acting. Me? I was a little more impulsive.

I walked over and sat next to her, plopping down on the bench with a sigh. "You know, if Zoe gets any more intense, I'm going to have to start carrying a fire extinguisher," I said, trying to keep it light even though I was seriously worried.

Xane smiled softly but didn't say anything right away, which wasn't surprising. She was always careful with her words. I waited for her to respond, knowing she'd say something thoughtful.

Finally, she glanced over at me. "It's getting worse, isn't it?" she asked quietly.

"Yeah," I said, nodding. "It's like she's spiraling. I mean, Zoe's always been confident, but now… I don't know. It's like she thinks she has to carry the whole band on her back or something. Like we're all just along for the ride."

Xane was silent again, but I could see the wheels turning in her head. "I think she's afraid of losing control," she said after a while. "Her pride is making her feel like she has to prove she's the best, and that's pushing everyone else away."

I leaned back on the bench, staring up at the darkening sky. "It's hard to watch, you know? She's been my friend for so long, but now it's like I don't even know how to talk to her. Everything sets her off."

Xane nodded, her expression serious. "Pride can be dangerous when it goes unchecked. It isolates people, makes them feel like they're the only ones who matter, the only ones carrying the weight. Zoe's feeling that pressure, and instead of leaning on us, she's shutting us out."

I ran a hand through my hair, frustrated. "I tried joking with her earlier, but that just made her mad. I don't know how to help her."

Xane turned to look at me, her gaze steady. "Maybe it's not about helping her right now. Maybe it's about addressing what's really going on—before it gets worse."

That made me pause. Xane wasn't wrong. We'd all been walking on eggshells around Zoe, trying to manage her mood without really confronting her about it. But avoiding the issue wasn't going to make it go away.

"Yeah, I guess you're right," I muttered. "But I'm not exactly the best at, you know, serious talks. I'm more of the 'make people laugh' type."

Xane smiled softly. "You don't have to be the one to have the serious talk. But maybe we all need to come together and remind Zoe that she's not alone. We're a band, Ruby. We do this together."

I let out a breath I didn't realize I was holding. Xane had a way of making things sound so simple, even when they weren't. But she was right. Zoe needed to hear from all of us—Leila, Lena, me, and Xane—that we were a team. That she didn't need to carry the weight alone.

"I just don't want to push her further away," I said, my voice quieter now. "She's my friend, you know? I don't want her to feel like we're ganging up on her."

Xane shook her head. "It's not about ganging up on her. It's about reminding her what we're about. Zoe's lost sight of the bigger picture, and it's up to us to help her find it again."

I nodded, feeling a little better. "Yeah, okay. I think you're right. We just need to figure out how to do that without making things worse."

Xane smiled at me, and I could see that glimmer of understanding in her eyes. "We'll find a way. We always do."

I chuckled, feeling a bit of the tension ease out of my chest. "You're good at this, you know? The whole 'calm and wise' thing."

Xane laughed softly, shaking her head. "I just try to listen."

"Well, you're doing a good job of it," I said, nudging her shoulder. "Thanks, Xane."

She gave me a small nod, and we sat there for a while, watching the last light of the sun fade behind the horizon. The problem with Zoe wasn't going to be solved overnight, but at least now I knew we'd face it together.

And with Xane's guidance, I had a feeling we might just be able to reach her before her pride tore the band apart.

13. Lena

It started like any other day on set—chaotic, stressful, and way too loud. But beneath the surface, I could feel the tension brewing. Zoe had been on edge all week, her frustration simmering just below the surface, and it seemed like no matter what anyone did, it wasn't enough.

We were halfway through another take, the cameras rolling as we played the same section of the song for what felt like the millionth time. I was trying to keep my head down, just doing my part and staying out of Zoe's way. But something in the air was off, and I could feel it creeping toward us.

"Cut!" Victor's voice echoed across the set, his frustration palpable. "Zoe, can you ease up on the drums? You're overpowering the track."

Zoe didn't even look at him, just stared straight ahead, her jaw clenched. "I'm playing it exactly how it needs to be played."

Leila tried to step in, her voice calm as always. "Let's just take it down a notch, Zoe. We're all a little tired."

Zoe turned to her, her eyes flashing with irritation. "Maybe if everyone else was playing with some actual energy, I wouldn't have to carry the whole thing."

I flinched at the edge in her voice. She wasn't just angry—she was blaming us. And I knew what was coming next before it even happened.

"Especially you, Lena," Zoe snapped, her eyes locking onto me. "You're barely even playing. I can't keep adjusting to your weak keyboard lines. You're dragging us down."

The words hit me like a slap, and for a moment, I couldn't move. Couldn't speak. My fingers hovered over the keys, trembling, as I tried to process what she'd just said.

I'd always been quiet, always stayed in the background. But Zoe's words cut deep, her accusation like a sharp blade. And for the first time, I felt something snap inside me.

"I'm not dragging anyone down," I said, my voice shaking but steady. "Maybe if you stopped trying to control everything, we could actually play as a band instead of catering to your ego."

Zoe's eyes narrowed, her expression turning cold. "My ego? Please. I'm the reason we're even on this set. You should be thanking me, not holding us back with your weak playing."

I could feel everyone's eyes on us, the tension so thick it was suffocating. Leila looked torn, Ruby's eyes were wide with concern, and Xane—Xane was watching closely, her face calm but focused, like she was trying to understand every word.

But I didn't care. Not anymore.

"You think you're carrying this band?" I said, standing up from the keyboard, my hands shaking with a mix of anger and hurt. "You think we're just here to make you look good?"

Zoe stood up from her drum kit, her arms crossed, her expression hard. "I'm the one keeping the rhythm, Lena. Without me, we'd fall apart."

The words stung more than I wanted to admit. But I wasn't going to let her walk all over me anymore.

"Maybe you should look around," I shot back, my voice rising. "Because we're already falling apart. And it's because of you."

Zoe's eyes flashed with something—surprise, maybe, or anger—but before she could respond, I grabbed my bag and stormed off the set, not even caring about the stunned silence that followed.

I could hear Ruby calling after me, and Leila trying to calm everyone down, but I kept walking, my heart pounding in my chest. The adrenaline was rushing through me, and I could barely think straight. All I knew was that I couldn’t stay there, couldn’t stand being *belittled* by someone I’d once considered a friend.

I found myself outside, the cool air hitting my face as I tried to catch my breath. My hands were still shaking, and I felt the sting of tears welling up in my eyes. But I didn’t want to cry. Not because of Zoe. Not after everything.

I had always been quiet. Always kept to myself. But that didn’t mean I was weak. And it didn’t mean I didn’t matter. I was part of this band—whether Zoe liked it or not.

As I stood there, trying to pull myself together, I heard the door creak open behind me. I didn’t have to turn around to know who it was.

"Lena," Xane’s voice was soft, calm, just like always. "Are you okay?"

I wiped at my eyes quickly, trying to compose myself. "No," I admitted, my voice barely a whisper. "I’m not."

She came to stand beside me, not saying anything for a moment, just letting the silence settle between us. But I knew Xane. She was always thinking, always analyzing, trying to figure out the best way to make sense of the situation.

"You stood up for yourself," she said after a while, her voice gentle. "That takes strength."

I shook my head, feeling a lump form in my throat. "It doesn't feel like strength. It just feels like I can't do this anymore."

"You're stronger than you think," Xane said quietly. "And Zoe… she's struggling with her own honor right now. It's clouding her judgment. But that doesn't mean what she said is true."

I let out a shaky breath, feeling the weight of her words sink in. "I don't know if I can keep doing this, Xane. She doesn't even care about the rest of us anymore. She just cares about herself."

Xane didn't deny it, but I could see the understanding in her eyes. "We'll figure it out," she promised. "But don't let her words make you doubt yourself. You belong in this band. We all do."

I nodded, feeling the tears start to slip down my cheeks despite my best efforts to hold them back. I wanted to believe her. I really did. But after everything that had happened, I wasn't sure if things could ever go back to the way they were.

Because once pride tears you apart, it's hard to put the pieces back together again.

And I wasn't sure Zoe even wanted to try.

14. Xane

After Lena stormed off the set, I could feel the weight of Zoe's words still hanging in the air. The band was shaken, and Zoe… she was standing by her drum kit, arms crossed, her face a mask of frustration. But I could see it now—this wasn't just about anger or even ego. This was pride, the kind that digs in deep and refuses to let go, isolating the person who holds it so tightly.

I had been watching Zoe closely for weeks, trying to understand what was driving her. It was clear she was pushing us all away, but now I knew why. Her pride had turned *toxic*, and she couldn't see the damage it was doing—to her, to us, and to the band.

As I stepped away from the chaos of the set, the familiar voice of the Analyst filled my mind, as calm and steady as ever. *"Pride is a dangerous thing, Xane. It can build walls around someone, making them feel invincible. But in the end, it only isolates them."*

I knew she was right. Zoe had always been proud of her talent, and rightfully so—she was one of the best drummers I'd ever seen. But somewhere along the way, that pride had shifted. It wasn't healthy anymore. It was making her believe she didn't need us, that she was better than the rest of us, and it was tearing the band apart.

"What do I do?" I asked quietly, my eyes drifting toward the horizon as I sought the answer. "How do I reach her?"

The Analyst paused for a moment, letting the question hang in the air before answering. *"Pride can lead someone down a path of isolation and disgrace if left unchecked. Proverbs 11:2 reminds us,* ***'When pride comes, then comes disgrace, but with humility comes wisdom.'*** *Zoe has lost sight of that humility, Xane. She's forgotten that true strength comes not from proving herself, but from working with others."*

The words resonated with me, settling into my thoughts like a piece of a puzzle clicking into place. Zoe's pride wasn't just about her drumming or the band. It was about her need to control, to protect herself from the fear of failure. But in doing so, she was alienating everyone around her.

"Pride isolates," I said softly, more to myself than to the Analyst. "It makes you feel like you're the only one who matters, like no one else can help you."

"Yes," the Analyst agreed. *"And that's why it's so dangerous. When pride blinds you, it keeps you from seeing the value of others. Zoe has lost her humility, and with it, she's losing the wisdom that comes from collaboration, from knowing she doesn't have to do everything alone."*

I thought about the tension between Zoe and Lena, the way Zoe had torn into her during rehearsal, belittling her contribution to the band. It wasn't just about the music. It was about Zoe's fear of not being in control. She was letting her pride take over, and it was driving a wedge between all of us.

"She's pushing us away," I murmured, feeling the weight of the situation. "She thinks she has to carry it all herself."

"Exactly," the Analyst responded. *"Pride makes people believe they can't rely on others, that asking for help is a sign of weakness. But true wisdom comes from recognizing the strength in humility, in knowing that you're part of something bigger than yourself."*

I knew that Zoe wasn't ready to hear this yet. She was too deep in her own head, too wrapped up in proving herself. But if she didn't let go of that pride soon, it would consume her—and the band with it.

"How do I reach her?" I asked again, feeling the urgency in my own voice.

"You can't force humility on someone," the Analyst said softly. *"But you can lead by example. Show Zoe that there's strength in being vulnerable, in relying on others. Remind her of what she once valued in the band—connection, teamwork, and shared success."*

I nodded, taking in her words. I didn't know how I was going to get through to Zoe, but I knew one thing for sure: I wasn't going to give up on her. Not yet.

Pride may have taken hold of her, but I had seen the real Zoe underneath—the one who cared about us, who cared about the music we made together. And if there was even the smallest chance that I could help her find her way back to that, I was going to take it.

"Thank you," I said quietly, feeling a sense of calm settle over me. The path ahead wasn't clear, but I knew what needed to be done.

As I turned back toward the set, I thought about Proverbs 11:2 again, letting its message guide me. Humility brings wisdom, I reminded myself. And maybe, just maybe, if I could show Zoe that humility wasn't weakness, she'd see the wisdom in it too.

I took a deep breath and made my way toward Zoe, ready to face whatever came next.

15. Xane

I knew this conversation wasn't going to be easy, but it was necessary. Zoe's behavior had been pushing everyone to their limits—especially Lena—and if I didn't say something now, I wasn't sure how much longer the band could hold on. Her pride was driving a wedge between all of us, and it was time to confront it head-on.

I found Zoe backstage, sitting alone with her drumsticks in hand, tapping them rhythmically against her leg. Her face was a mask of frustration, but beneath it, I could see the exhaustion, the tension that had been building for weeks. She looked up when she heard me approach but didn't say anything, just kept tapping the sticks in a steady beat.

"Zoe," I started, keeping my voice calm but firm. "We need to talk."

She sighed, her shoulders slumping slightly. "Look, Xane, I don't want another lecture, okay? I'm just trying to get through this."

I took a deep breath, trying to steady myself. "I'm not here to lecture you. I'm here because I care about you—and about the band."

Zoe's eyes flickered with something—defensiveness, maybe—but she didn't interrupt, so I pressed on.

"I know things have been tough," I said, sitting down next to her. "The pressure is getting to all of us. But your pride, Zoe… it's hurting the band. It's hurting Lena."

Her jaw tightened, and I could see the walls going up. "I'm not the problem, Xane. I'm just trying to keep things together. If anything, Lena needs to step up. We can't keep holding her hand."

"That's not fair," I said softly but firmly. "Lena's doing her best, just like the rest of us. But you're shutting her out. You're shutting all of us out."

Zoe's hands stilled, the tapping of her drumsticks ceasing. She looked away, her expression hardening. "I'm not shutting anyone out. I'm just trying to make sure we don't fall apart."

I shook my head, my heart heavy. "But don't you see? That's exactly what's happening. The more you try to control everything, the more it slips away. Your pride is blinding you, Zoe. You're not listening to us anymore. You're not seeing us as a team."

She flinched at that, but her defenses were still up. "I'm doing what I have to do. Someone has to keep things moving."

I sighed, feeling the weight of this moment pressing down on me. "We all want the same thing, Zoe. We all want the band to succeed. But this—what's happening between you and Lena—it's not just about the music. It's about trust. And right now, we're losing that because you're letting your pride get in the way."

Zoe's face twisted in frustration. "You think I'm proud? That I'm doing this because I think I'm better than everyone else?"

I looked her in the eyes, steady and unflinching. "Yes. And no. I think you're proud of your talent, and you should be. But I also think you're scared. Scared of failing, of losing control. And that fear is turning into pride—the kind of pride that makes you feel like you have to prove something, like you have to carry all of this on your own."

Her eyes widened slightly, and for the first time, I saw a crack in her armor.

"You don't have to do it alone, Zoe," I said softly. "That's the thing about being in a band. We're in this together. But right now, you're pushing us away."

She was quiet for a long time, her eyes downcast as she gripped the drumsticks tighter in her hands. I could see the battle going on inside her—the conflict between the walls she'd built and the truth she didn't want to face.

Finally, she spoke, her voice barely above a whisper. “I don’t know how to let go. I’ve always been the one keeping things steady. If I let go, what happens?”

I reached out, gently placing a hand on her arm. “You don’t have to hold everything so tightly, Zoe. That’s not what makes you strong. What makes you strong is your ability to trust us, to trust that we’re all in this together. Pride can give you confidence, but too much of it—when it turns into arrogance—isolates you. It’s not about proving yourself to us. We already know how talented you are. But you need to trust that we’re talented too. We need you, Zoe, but we need all of us.”

Zoe’s grip on the drumsticks loosened, and she let out a shaky breath. “I didn’t mean to hurt Lena. I just… I don’t know how to deal with all of this.”

I gave her a small, encouraging smile. “We can figure it out. But the first step is acknowledging that you don’t have to carry the whole band by yourself. Let us help you. We’re stronger together than we are apart.”

She looked up at me, her defenses finally crumbling. “I’m sorry, Xane. I didn’t realize…”

“It’s okay,” I said gently. “But we need to fix this. Talk to Lena. Make things right. It’s not too late.”

Zoe nodded slowly, the weight of her pride visibly lifting, even if just a little. “I will. I’ll talk to her.”

As she stood up, ready to make amends, I felt a sense of relief wash over me. This was just the beginning, but it was a step in the right direction.

The Analyst’s words echoed softly in my mind, reminding me of the lesson we were all learning: “When pride comes, then comes disgrace, but with humility comes wisdom.”

Zoe was finding her way back to humility, and with it, she was finding her way back to the band.

16. Zoe

Xane's words stuck with me, echoing in my mind long after she walked away. I hadn't expected her to confront me like that, to call me out so directly. And yet, she hadn't been harsh. She hadn't been angry. She had been… calm. Steady. Exactly what I wasn't.

Sitting alone with my drumsticks in hand, I found myself replaying the last few weeks, trying to understand where everything had gone wrong. When had my pride started turning into something ugly? Something that hurt the people I cared about?

I glanced around the empty set, my mind going back to the diva actors we'd been dealing with during this shoot. I had rolled my eyes at them—at their ridiculous demands, their superiority complex, the way they treated everyone else like they were beneath them. I'd seen how their arrogance drove wedges between them and everyone around them. I despised it.

And yet, Xane's words kept coming back to me, like a mirror I didn't want to look into. You're letting your pride isolate you.

Had I become the very thing I hated?

I thought about Lena, about how I'd snapped at her earlier. I had been so focused on getting everything perfect, on making sure the band didn't fall apart under the pressure of this music video, that I hadn't even realized how harsh I'd been. I told myself I was doing it for the good of the band, but deep down, I knew the truth.

I wasn't trying to protect the band. I was trying to protect me.

I was afraid. Afraid of failure, of not being good enough. And instead of facing that fear, I let it turn into something worse. Pride. The kind of pride that makes you think you don't need anyone. The kind that blinds you to the people standing right next to you.

I closed my eyes, the weight of it all pressing down on me. Xane had been right. My pride wasn't just confidence anymore. It was arrogance. And it was hurting the people who mattered most.

I didn't want to be like those diva actors, pushing everyone away just to protect my own ego. I didn't want to be the reason the band fell apart. But if I kept going down this path, that's exactly where I was headed.

When pride comes, then comes disgrace, Xane had said, quoting some Bible verse I'd heard before but never really thought about. But with humility comes wisdom.

Humility. That's what I had lost. Somewhere along the way, I'd stopped seeing the band as a team and started seeing them as an extension of my own success. I had been so caught up in proving myself, in being the best, that I forgot why we started this band in the first place.

It wasn't about me. It was about us. About making music that mattered, music that we believed in. And that was something I couldn't do alone.

I let out a long breath, the realization settling in my chest. I had to fix this. I had to talk to Lena. To the whole band. I had to make things right, not just for them, but for me.

Because if I kept letting my pride control me, I wasn't just going to lose the band. I was going to lose myself.

I stood up slowly, the weight of my drumsticks feeling heavier than they ever had before. But this time, it wasn't just the burden of the music. It was the weight of responsibility—the responsibility to be better, to do better. Not just for the band, but for myself.

I wasn't the diva. I wasn't the arrogant actor who thought they were above everyone else. I wasn't going to let my pride destroy everything I'd worked for.

But I had to start with an apology. And that… that was going to take more strength than I had shown in a long time.

I tightened my grip on the drumsticks and headed toward the door, ready to face the consequences of my actions. Ready to let go of the pride that had been holding me back.

Because Xane was right. Pride can isolate you, but humility—humility brings wisdom. And if I wanted to keep this band together, if I wanted to be the kind of person I respected, I needed to find that humility.

Starting now.

17. Leila

The second Victor announced the delay, I could feel the tension in the air spike. The production crew was scrambling around, talking in low voices, trying to figure out how to salvage the mess. Apparently, some key piece of equipment had malfunctioned—again—and now the entire shoot was delayed indefinitely.

It wasn't like we hadn't been through delays before, but this one hit differently. We were already on edge, and this felt like the final straw for everyone.

I looked around at the band. Ruby was chatting with one of the sound engineers, cracking jokes like she always did to keep the mood light, but even she seemed a little deflated. Xane stood off to the side, lost in thought, no doubt reflecting on everything that had gone down with Zoe and Lena. And then there were Zoe and Lena themselves—each in their own corner, barely looking at each other.

The tension between them had been lingering ever since that argument on set. Lena had stormed off, and even though Zoe hadn't said much since, I could tell she was stewing over it. Normally, I'd jump in to smooth things over, but this time… I wasn't sure how. Things felt fragile, and I was afraid that one wrong word might shatter whatever fragile peace we had left.

"Hey, Leila." Ruby's voice broke through my thoughts as she plopped down beside me. "Another delay. Yay us."

I smiled faintly, grateful for her attempt to keep things light, even if it wasn't working. "Yeah. Just what we needed."

Ruby glanced around the set, then leaned in closer, lowering her voice. "So, how long are we going to let the whole Zoe-and-Lena situation simmer before it boils over?"

I sighed, running a hand through my hair. "I don't know. I've been trying to figure out how to handle it, but they're both so... on edge."

Ruby nodded, crossing her arms as she glanced over at Lena, who was tinkering with her keyboard in silence. "Yeah, it's bad. I don't think I've ever seen Lena this upset."

"And Zoe's not exactly helping," I muttered, my eyes shifting toward Zoe, who was sitting behind her drum kit, looking like she was ready to snap at the next person who spoke to her.

"I thought Xane talked to her?" Ruby asked, raising an eyebrow.

"She did," I said, rubbing the back of my neck. "But I think it's going to take more than one conversation to fix this."

Ruby let out a low whistle. “Yikes. Well, in the meantime, we’re going to have to do something, or this whole thing is going to blow up before we finish the shoot.”

I nodded, knowing she was right. We couldn’t afford to let this tension fester any longer. The band needed to get back on track, and it was up to me to hold us together, just like always.

Taking a deep breath, I stood up and made my way toward Zoe. She looked up as I approached, her expression guarded, but I could see the frustration in her eyes.

“Hey,” I said softly, sitting down beside her. “How are you holding up?”

She shrugged, tapping her drumsticks against her leg. “I’m fine. Just waiting for this whole thing to get moving again.”

I studied her for a moment, trying to gauge how much I could push. Zoe had always been a bit of a perfectionist, but lately, her pride had been getting the best of her, especially when it came to Lena. But beneath that pride, I could see the exhaustion. She was tired—tired of trying to hold everything together, tired of the delays, and, I suspected, tired of fighting with Lena.

"Look," I said carefully, "I know things have been tense lately. And I get that you're frustrated. But we've got to work through this together."

Zoe didn't say anything, just kept tapping her sticks against her knee. I waited, letting the silence stretch between us, hoping she'd open up.

Finally, she sighed, her shoulders slumping slightly. "I know, Leila. I know I messed up with Lena, okay? I'm just… I'm trying to figure it out."

I nodded, feeling a flicker of relief that she was willing to admit it. "That's a start. But we can't fix this if you and Lena are still at each other's throats. We're a band, Zoe. We've got to stick together, especially now."

She looked over at Lena, her jaw tightening. "I don't know how to fix it. I was just trying to keep everything from falling apart."

"I know," I said gently. "But you can't do it alone. None of us can."

Zoe was quiet for a long moment, then she nodded, the tension in her posture easing just a little. "I'll talk to her. I'll try to make it right."

"Good," I said, standing up. "Because we need you both. The band doesn't work without all of us."

I walked away, feeling a little lighter but knowing there was still more work to be done. I made my way over to Lena, who was sitting by her keyboard, looking more withdrawn than I'd ever seen her.

"Hey," I said, sitting down beside her. "You okay?"

Lena shrugged, not looking up from the keys. "Just... thinking."

"About Zoe?" I asked, keeping my voice soft.

She hesitated, then nodded. "Yeah. I don't know, Leila. I don't know if we can get past this."

I reached out and placed a hand on her shoulder. "We will. Zoe knows she messed up. She's going to talk to you. But you've got to be willing to hear her out."

Lena finally looked up, her eyes filled with doubt. "What if things don't go back to the way they were?"

"They won't," I said honestly. "But maybe that's a good thing. Maybe we'll come out of this stronger. As long as we're willing to try."

Lena gave a small nod, but I could tell it would take more time for her to fully believe that. Still, it was a step in the right direction.

As I stood up, I took a deep breath, feeling the weight of responsibility settle on my shoulders. It wasn't going to be easy, but we'd been through tough times before. And we'd come through it together.

I just had to keep holding the line, no matter how long it took.

18. Lena

I sat at the keyboard, my fingers resting lightly on the keys, but I wasn't playing. I hadn't played much since the argument with Zoe. Every time I tried to lose myself in the music, her words echoed in my mind—words that made me feel small, replaceable.

We could get anyone to play those keyboard parts.

It had stung more than I'd let on. I'd always been the quiet one, the one who didn't need the spotlight. But that didn't mean I didn't care. I cared deeply about the band, about my role in it. And for Zoe to dismiss my contribution like that… it hurt. It made me question everything.

But as I sat there, staring at the familiar black and white keys, something shifted. I wasn't angry anymore. I wasn't even hurt—not really. Instead, I felt something new, something I hadn't allowed myself to feel in a long time: pride.

Not the kind of pride Zoe had—arrogant, defensive, always needing to prove herself. No, this was different. It was quieter, deeper. It was the kind of pride that came from knowing who I was and what I brought to the band.

I didn't need Zoe's approval. I didn't need anyone's approval, really. I knew my worth, and I knew what I added to our music. My playing wasn't flashy or loud, but it had its place. It was part of the heartbeat of the band, the rhythm that held everything together, even if it wasn't always front and center.

I took a deep breath and closed my eyes, letting my fingers find their way across the keys. The sound was soft, almost delicate, but it filled the space around me, wrapping me in a sense of calm. For the first time in weeks, I wasn't thinking about Zoe's words. I wasn't thinking about the tension or the arguments. I was just playing, letting the music flow through me, reminding myself of why I loved it in the first place.

As the melody grew, so did my confidence. I had spent too long doubting myself, too long letting someone else's opinion cloud my own sense of worth. But now, in this moment, I realized that I didn't need validation from anyone but myself.

I was good at what I did. Maybe I wasn't the loudest or the most attention-grabbing, but I didn't need to be. My music spoke for itself.

A small smile crept onto my face as I played, the notes becoming stronger, more assured. This was who I was—quiet, thoughtful, but powerful in my own way. I didn't need to prove anything to anyone, least of all Zoe.

I let the last note linger in the air, feeling the resonance in my chest. It felt like a release, like I was letting go of all the doubt and fear that had been weighing me down. I wasn't going to let anyone make me feel small again. Not Zoe, not anyone.

As I stood up from the keyboard, I felt lighter, more grounded in who I was. I didn't need to fight for my place in the band. I had earned it. And no one—no matter how frustrated or prideful—could take that away from me.

From now on, I would play for me. For the love of the music. And if that wasn't enough for someone else, that was their problem, not mine.

I was proud of what I brought to the band. And that was enough.

19. Xane

The tension in the air had become almost unbearable. It wasn't the kind of pressure you wanted before a big music video shoot—the kind that pushes you to perform harder, to get everything right. No, this was the silent kind. The kind that sticks to the walls, clings to conversations, and makes it feel like everything could fall apart at any moment. And most of it was thanks to Zoe.

Zoe had always been intense, but over the last few weeks, she'd been dialing it up to eleven. I could see it weighing on everyone, especially Lena. Zoe's arrogance wasn't new, but lately, it felt like she was steamrolling through every rehearsal, every shoot, like the rest of us were just set pieces in her grand show. Lena, quiet and thoughtful as ever, was starting to retreat further into the background. I didn't blame her. Zoe had a way of making you feel like you were unnecessary, even when you knew you weren't.

I caught Ruby's eye across the room. She was watching the same thing I was—Zoe, taking command of the set like a general, her voice cutting through every conversation, every plan. Lena stood awkwardly near the back, arms crossed, her face tight with a familiar look of unease. Ruby rolled her eyes at me. I couldn't help but smirk, but beneath the humor, we both knew something had to change.

During a quick break in filming, I walked over to where Ruby was leaning against a speaker.

"She's gonna snap at someone any minute," I muttered, just loud enough for Ruby to hear.

Ruby sighed, rubbing the back of her neck. "Tell me about it. I'm this close to snapping back at her myself."

"Lena's shutting down. You can see it."

Ruby nodded. "She's barely said two words all day. It's not good. Zoe's in overdrive, and it's killing the vibe."

We stood there for a moment, the hum of the crew around us filling the silence between us. This wasn't what we wanted. We needed to be united. When the music video dropped, it would be all of us, not just Zoe, that people were watching. But right now, the band felt like it was fracturing. Ruby glanced at me, a thoughtful expression crossing her face.

"Maybe we need a reset. Something to break up the tension," she said.

I raised an eyebrow. "Like what?"

Ruby grinned. "I was thinking—what if we take the band to an escape room?"

I blinked. "An escape room? Seriously?"

“Hear me out,” Ruby continued, pushing away from the speaker and gesturing with her hands like she was pitching a brilliant new idea. “Think about it. We’re all stuck in a room, we have to solve puzzles, work together, and there’s no way Zoe can just boss everyone around. It’s the perfect way to get us all back on the same page.”

I crossed my arms, considering it. Ruby wasn’t wrong. An escape room might force Zoe to slow down and actually listen to someone else for once. And Lena… well, maybe it would give her a chance to step up, to feel like she was contributing more than just standing in the background.

“I don’t know if Zoe’s gonna be thrilled about it,” I said cautiously.

Ruby laughed. “She’s not gonna be thrilled about anything that’s not her idea. But once we’re in there, she’ll have to go along with it.”

“And Lena?”

“Lena will love it. It’ll give her something to do, make her feel part of the team again. Plus, she’s smart as a whip. She’ll figure stuff out that the rest of us wouldn’t even see.”

I nodded, starting to see the merit in the plan. “Okay. Let’s do it. But how do we convince everyone to actually go?”

Ruby smirked. “Leave that to me.”

Later that afternoon, we gathered the band together after filming wrapped for the day. Zoe, looking tired but still as fiery as ever, flopped down on a couch while Lena hovered nearby, clearly unsure whether she should sit or stand.

“So,” Ruby began, clapping her hands together, “we’ve been working our rear ends off, right? And I think we need a little break. Something fun. Something to get us out of our heads.”

Zoe raised an eyebrow. “And what exactly do you have in mind?”

Ruby grinned. “An escape room.”

Zoe blinked, then snorted. “An escape room? What are we, teenagers?”

“No,” Ruby said, undeterred, “we’re a band that could use a little team-building. Think about it—getting locked in a room, solving puzzles, using our brains instead of bashing our heads against walls all day. We’ll have fun, and it’ll help us work together better.”

I chimed in, backing Ruby up. “We’ve been grinding non-stop. A break could do us all some good. Plus, it'll be a way to get our heads clear before the next shoot."

Lena looked intrigued but stayed quiet, watching Zoe closely. Zoe, on the other hand, seemed more skeptical.

"I don't know..." Zoe started, but Ruby cut her off.

"Come on, Zoe, don't tell me you're scared you won't be able to solve the puzzles," Ruby teased, her grin widening.

Zoe shot her a glare, then sat up straighter. "Please, I could solve it in my sleep."

"Then let's go prove it," Ruby said with a wink.

Zoe sighed dramatically, but a small smirk tugged at the corner of her mouth. "Fine. I'm in."

I looked over at Lena, who was still hesitating. "What about you, Lena? You down?"

Lena finally smiled, her shoulders relaxing a bit. "Yeah. I think it could be fun."

With everyone on board, I felt a small surge of relief. Maybe Ruby was right—this might just be what we needed to break the tension and remind everyone we're a team. Zoe would have to check her ego at the door, and Lena would get her chance to show her worth.

As we headed out to book the escape room, I couldn't help but feel hopeful. This was exactly the kind of reset we needed.

20. Xane

The moment the heavy iron door clanged shut behind us, the lights flickered dimly, casting long shadows across the stone walls. The escape room looked like something out of a medieval dungeon—chains, rusted metal bars, and cobwebs clinging to the corners. It was part of the theme, but the atmosphere was definitely unsettling. I could sense the unease creeping in, even though none of us said it aloud.

"Well, this is cozy," Ruby muttered, trying to lighten the mood as she glanced around. Leila chuckled softly from where she stood, arms crossed, but her sharp eyes were scanning the room, already figuring out the layout.

It wasn't long before Zoe stepped forward, taking control like she always did. "Alright, let's get moving. Spread out and look for clues. Anything that stands out could be useful—a key, a code, whatever."

She gave directions like she was organizing the next take of our music video, completely in command. I shot a glance at Leila, who gave me a small nod. We both knew this was exactly what we wanted to avoid—Zoe steamrolling through everything while the rest of us tried to keep up. But Leila didn't step in yet. She was watching, letting it play out for now.

Meanwhile, Lena was standing near the back, her shoulders hunched as she hesitated to join in. I could see she wasn't sure where she fit into this, just like how things had been in the band lately. Ruby caught my eye, and we exchanged a knowing look—this was Lena's moment to find her place, to show that she was just as valuable as anyone else here.

"Anything jump out at you?" I asked Lena quietly, walking over to her.

She glanced around nervously before nodding toward a section of the wall. "Those bricks over there... they look different. Maybe there's something behind them?"

I smiled. "Good catch. Let's check it out."

As Lena and I worked on the wall, Zoe marched through the room, yanking on chains and pulling random objects off shelves. Ruby was crouched near a chest, fiddling with the lock, while Leila, ever the observant one, was studying an old map hung on the wall.

"We need to focus," Zoe snapped, her voice sharp. "You're all over the place. We don't have time to waste."

Ruby looked up, raising an eyebrow. "We're not all over the place—we're checking the room. Relax."

Zoe's frustration was building. "We're supposed to be working together, not just wandering around."

"Exactly," I said, pulling a brick loose to reveal a hidden compartment. Inside was a rolled-up parchment. I carefully unrolled it, revealing a riddle. "'In the darkest hour, you must find the light. Three flames must burn, but one must ignite.'"

Zoe groaned. "Of course, a riddle. Great."

Leila stepped forward, studying the words over my shoulder. "Three flames, huh? We should look for something related to fire. Candles, torches—there's got to be something like that in here."

Lena's quiet voice cut through the chatter. "Maybe it's not just about actual flames. It could be about light or something symbolic."

Zoe, still on edge, waved off the suggestion. "We don't have time for symbolism, Lena. We need to solve this and move on."

Lena looked down, biting her lip, but before Zoe could charge ahead, Leila stepped in smoothly. "Zoe, take a breath. We're all working toward the same goal here, but we've got to think things through. If we rush it, we'll miss something important."

Zoe clenched her jaw but didn't argue back. Leila had that kind of influence—her words carried weight because she wasn't just the lead vocal or guitarist.

She was the band's manager, the one who kept everything on track, and even Zoe knew not to push too far when Leila spoke.

"Alright, so let's start with the basics," Leila continued. "We've got three flames to light. There's a candelabra here with three holders—two candles, but one's missing. What do you think?"

Ruby held the candelabra up. "I think we need to figure out what that third flame is. We've got two real candles, but maybe Lena's right—it's more than just fire."

Lena looked surprised that Ruby had backed her up but nodded slowly. "Maybe it's a torch or something? I saw a holder on the far wall that might fit."

We all moved toward the holder Lena had spotted, and sure enough, there was an unlit torch resting inside. I grabbed it and used one of the candles to light it. As soon as the flame caught, a faint clicking sound echoed through the room, and a hidden compartment in the floor opened, revealing a small key.

"That's it!" Ruby said with a grin, turning to Lena. "Nice call."

Lena smiled, but it was tentative—like she wasn't quite used to the idea that her suggestions were actually valuable. I could see the confidence starting to build in her, though. And for the first time, Zoe didn't try to jump in and take credit for the solution. She just grabbed the key and moved toward the next door.

Leila held the door open as we all stepped into the next chamber, her calm energy keeping everyone balanced. She had a knack for smoothing out the edges when things got tense. Zoe might be the loudest, but Leila was the one with real control.

As we entered the next room, I could feel something shifting. Zoe wasn't running the show anymore, and Lena was beginning to step out of the background. Ruby was right—this escape room was giving us what we needed. A chance to stop and see each other's strengths in a whole new way.

But we weren't out of the dungeon yet, and the puzzles were only going to get harder from here.

21. Xane

The next room was by far the most complicated we'd seen yet. Dim light flickered across five strange-looking mechanisms scattered throughout the space. Each one was connected by a series of ropes, gears, and pulleys that crisscrossed the stone walls like some kind of medieval machine. A plaque near the door read, **"All must move as one."**

Zoe, of course, stepped forward first, her eyes darting from one mechanism to the next, sizing up the challenge like it was another performance she had to command. "Okay, let's break this down. I can handle two of these myself—pull the lever over there and shine the laser here. Someone else can handle the easy stuff, like rolling the ball."

Leila frowned, stepping forward to study the setup. "Hold on, Zoe. It says everything needs to happen at the same time. We can't rush this or do it alone. We all have to take a part."

Ruby wandered toward the center of the room, examining the mechanisms closely. "Yeah, Zoe, I think Leila's right. The timing has to be perfect, or nothing's going to happen. This is a team thing, not a one-person show."

Zoe rolled her eyes. “I know, but if I take on the tricky parts, it’ll be faster. We don’t have time to mess around.”

I could see where this was going—Zoe, trying to control everything again, thinking she could push us through on her own. But this puzzle wasn’t going to be solved with brute force or speed. I stepped forward before Zoe could grab onto two tasks at once.

“It’s not about who does the most,” I said firmly. “It’s about everyone doing their part together. That’s the whole point of this. We each need to focus on one thing and do it right, or we’ll keep missing the mark.”

Zoe stared at me for a moment, her frustration simmering beneath the surface, but I wasn’t backing down. Slowly, she gave a small shrug, stepping back. “Fine. Let’s divide it up.”

Lena, who had been standing by the door, watching everything unfold, shifted nervously as the conversation unfolded. She hadn’t said much since we entered the room, and I could tell from the look on her face that she was questioning where she fit into all this. She was feeling the same way she did in the band—like her role wasn’t as important, that she didn’t have a major part to play.

Ruby noticed, too. She walked over to Lena, her voice quiet but encouraging. “Hey, this is your chance to show what you've got. We need you. None of this works without everyone pitching in, okay?"

Lena looked up, hesitating, but Ruby smiled warmly. "I know it's weird, but trust me, you've got this."

Slowly, Lena nodded, her confidence starting to surface. "Okay," she whispered. "I'll try."

We gathered around, each taking a mechanism. Ruby was by the laser, ready to aim it at the target across the room. Leila, with her steady hands, positioned herself at the rope. Zoe, somewhat reluctantly, moved to the lever, and Lena stood by the ball, which needed to roll down a narrow path into a small hole to trigger the final mechanism. I took the knob, ready to turn it at the exact moment everything else fell into place.

"All right, on three," Leila said, glancing around to make sure we were all ready. "One, two… three!"

At the same instant, we each moved. Zoe pulled the lever, Ruby fired the laser, Lena rolled the ball, Leila tugged the rope, and I twisted the knob. For a moment, nothing happened. My heart sank—had we missed it? But then, with a soft *clunk*, the gears around us began to turn, chains rattling and pulleys sliding into place. A slow rumble built from beneath the floor as the entire room seemed to shift.

The door creaked and groaned, and then—finally—it swung open.

We all stood there for a moment, almost not believing we had pulled it off. There was a strange, quiet satisfaction in the air—a sense of small victory, but a significant one.

"Nice work, team," Leila said with a smile, her calm voice carrying a hint of pride.

Ruby gave Lena a playful nudge. "See? Told you you had this."

Lena smiled shyly, but this time, there was no doubt in her expression. She'd done it. We'd all done it, and for the first time, it felt like Lena had truly stepped into her own. She hadn't just followed along; she had contributed something real, something crucial.

Zoe, standing next to the lever, took a breath and, surprisingly, didn't say a word about how she could have done more. She didn't claim the credit this time. Maybe it was because she knew, deep down, this was a group effort. Or maybe she was starting to realize that she couldn't always do everything on her own.

As we walked through the door and into the next chamber, the band felt a little more like a team again. Sure, we still had more puzzles ahead, but for the first time in a while, I felt like we were actually working together.

22. Xane

The door creaked open, revealing the final room of the escape. It was bigger than the others, with a high, vaulted ceiling and strange symbols painted on the walls. At the far end were two intricate contraptions, both connected by a web of ropes, pulleys, and gears. The whole room was divided into two distinct sides—one for each puzzle. A plaque on the wall read, **“Two paths, one goal.”**

I could feel the weight of this last challenge hanging over us. We’d come so far, but this was different. It was more complicated. We weren’t just relying on one person solving a puzzle at a time—this required us to split up, each team working separately but in perfect sync with the other.

Zoe was already on the move, walking to the center of the room, her eyes scanning the space like she was ready to take control again. “Okay, listen up. We’ll split into two groups. I’ll take Leila and Lena on one side, and Ruby, you and Xane handle the other.”

Before anyone could speak up, Zoe continued, “We’ll need to communicate constantly, and I’ll call the shots from my side. Let’s keep it tight, and we’ll get this done quickly.”

I could feel the old tension rising again. Zoe wasn’t asking for input—she was taking charge like usual.

And while we'd worked together better in the last few rooms, it was clear she still had that instinct to control everything.

Before she could finish laying out her plan, I stepped forward. "Zoe, this isn't about one person running the show. We need to trust each other, not just follow orders. If we're going to make this work, we have to listen to everyone."

Zoe stopped, her eyes narrowing slightly as she processed my words. I could tell she wasn't used to being challenged like this, but she didn't push back immediately. There was a flicker of understanding behind her confident exterior—maybe all the rooms before this one had started to sink in.

"Okay, fine," she said, though her voice was still edged with impatience. "But we need a plan."

I turned to the group. "Let's split up like Zoe suggested, but instead of just following one person's lead, we each take a piece of this. It's going to take everyone's input to make these puzzles work together. Ruby and I will handle one side, and Lena, you take the lead on the other with Zoe and Leila backing you up."

Lena's eyes widened slightly. "Me? Lead?"

I nodded, giving her a reassuring smile. “You’ve got a great eye for details, and this is the perfect chance to show it. We trust you.”

Zoe shifted, looking like she was about to protest, but then her gaze landed on Lena. For a moment, I saw something soften in Zoe’s expression—maybe it was the realization that she needed to give others space to step up. She didn’t argue. Instead, she simply nodded.

“Alright, Lena,” Zoe said, her voice calmer now. “We’ll follow your lead.”

Lena looked surprised, but there was also a spark of determination in her eyes. She glanced over at Leila, who gave her a supportive nod, and then at Ruby, who flashed her a grin.

“We can do this,” Lena said softly, more to herself than anyone else, but I could hear the confidence building.

With the teams set, we moved to our respective sides. The puzzle on our side involved a large, rotating mechanism and a series of colored levers that needed to be pulled in a specific sequence. Across the room, Lena’s team was dealing with a more complex arrangement of gears and locks that could only be unlocked if they mirrored our movements exactly. It was like trying to solve two puzzles that were wired together, and we couldn't succeed unless both sides worked in perfect harmony.

Ruby and I focused on the rotating mechanism, adjusting the levers according to the symbols that flashed above the gears. Every move we made needed to be coordinated with Lena's side. I glanced over at Lena, who was explaining something quietly to Zoe and Leila.

"Lena," I called across the room, "we're about to pull the second lever. Does that match with the gears on your end?"

She looked at the contraption in front of her and then at Zoe, giving her instructions to adjust a dial. "Almost. Give us a second… Okay, go for it!"

We pulled the lever, and the mechanism on our side rotated smoothly. A soft click echoed through the room as part of the puzzle unlocked. I could see a small smile form on Lena's face. She was in her element now, thinking things through, offering clear instructions. It was the most confident I'd seen her all day.

"Great job, Lena," Ruby said, adjusting the next lever on our side.

As we moved through the sequence, each team relied on the other more and more. There was no room for miscommunication—if one side didn't keep up, the whole puzzle would reset. But something was different this time. There was no tension, no one trying to overpower the other. Zoe wasn't barking orders or rushing ahead. She was listening, following Lena's lead, and making sure everything was in sync.

After what felt like hours but was probably only minutes, the final piece slid into place. The gears on both sides clicked into alignment, and a loud, satisfying *clang* signaled that the door was about to open.

We all stood still for a moment, just listening to the sound of the heavy door slowly creaking open. It wasn't just the sound of finishing the puzzle—it was the sound of us working together, really working together, for the first time in a long while.

As the door swung wide, we stepped out into the hallway, leaving the escape room behind. But something had shifted in the air. This wasn't just about solving a game; this was about understanding each other. Lena, once hesitant and unsure, had taken control and shown us her strengths. Zoe, always confident, had learned to step back and trust her bandmates. And the rest of us had found a way to work together without needing anyone to lead the charge.

We stood outside the exit, catching our breath. Zoe, who usually had something to say in moments like this, was quiet for once. Then she turned to Lena and gave her a genuine smile. "You did good, Lena. Real good."

Lena blushed slightly but smiled back, her head held higher than before. "Thanks. It wasn't just me, though."

Ruby threw her arm around Lena's shoulder. "Yeah, but you're a big part of why we got through that last one. Own it."

As we made our way back to the car, I glanced around at the band—Zoe, Ruby, Lena, and Leila. The escape room had forced us to rely on each other in ways we hadn't during rehearsals or even performances. It wasn't just about the music or the show anymore. It was about being a team.

We walked out of that escape room more connected than when we'd walked in. It felt like something had clicked into place, not just in the puzzles, but in us. And for the first time in a while, I was excited to see where we'd go from here.

23. Zoe

I had been dreading this moment for days, but I knew it was time. No more avoiding it, no more excuses. I owed Lena an apology. A real one.

I found her at the keyboard, her fingers gently tapping the keys in that way she always did when she was lost in thought. Her music was soft, almost like a whisper, but it carried weight. It always did. And for the first time, I really listened. I could hear how much she added to the band, how her playing filled the spaces between the rest of us, giving the music depth and meaning.

How had I missed that?

Taking a deep breath, I approached her, my steps feeling heavier than usual. I wasn't used to admitting I was wrong. But if I didn't fix this now, I knew the damage would only get worse.

"Lena?" I said softly, standing a few feet away, not wanting to interrupt her playing too abruptly.

She looked up, her fingers freezing over the keys, her expression cautious. "Hey."

The tension was still there, and I hated that I was the reason for it. But I couldn't blame her for being wary. I had hurt her, and she had every right to be distant.

“I… I wanted to talk,” I started, feeling the words get stuck in my throat. How was I supposed to do this? Apologies weren’t exactly my strong suit.

Lena nodded slowly, but she didn’t say anything. She was giving me the space to speak, and I knew I had to make this count.

“I’ve been thinking a lot about what I said to you,” I continued, my voice quieter than usual. “About how I’ve been acting.”

Lena’s eyes flickered with something—maybe surprise, maybe disbelief—but she still didn’t interrupt. I took that as a good sign.

“I was wrong,” I admitted, the words feeling heavy but necessary. “I shouldn’t have said those things. I shouldn’t have made you feel like you didn’t matter.”

She shifted slightly, her fingers resting lightly on the keys, but I could tell she was listening intently.

“I let my pride get in the way,” I confessed, finally facing the truth. “I was so focused on trying to keep everything perfect, trying to prove myself, that I pushed you away. I pushed all of you away. And that wasn’t fair.”

Lena’s expression softened just a little, but she still stayed quiet, letting me say what I needed to.

"I've always been proud of my drumming, of the band," I said, struggling to keep my voice steady. "But I forgot that it's not just about me. It's about us. And you… you're a huge part of that, Lena. I see that now."

Her eyes met mine, and for the first time in what felt like forever, I saw a flicker of understanding there. Maybe even hope.

"I'm really sorry," I said, my voice barely above a whisper now. "I never should've said you were replaceable. You're not. You never were."

Lena stared at me for a moment, her expression unreadable. I held my breath, waiting for her to say something—anything. I was ready for her to be angry, ready for her to tell me I didn't deserve forgiveness.

But instead, she surprised me.

"I appreciate that," she said softly, her voice steady but kind. "And I forgive you."

I let out a breath I didn't realize I had been holding. Relief washed over me, but it was more than that. It was like a weight had been lifted off my chest, a weight I had been carrying for far too long.

"I want us to move forward," Lena continued, her fingers brushing the keys again, but this time with more confidence. "But I need you to understand, Zoe, that it's not just about saying sorry. It's about really seeing me, seeing all of us, as equals."

I nodded, knowing she was right. "I do. I promise I do."

There was a long pause, and then, slowly, Lena gave me a small smile. It wasn't big or overly enthusiastic, but it was real. It was a start.

"We're a team," she said, echoing the words Xane had said to me earlier. "We're only as strong as we are together."

I smiled back, feeling something inside me soften—something I hadn't felt in a long time. Humility. I wasn't just drumming for myself anymore. I was drumming for us. For the band.

"Thanks, Lena," I said quietly, feeling more genuine than I had in weeks.

Lena nodded, then gently pressed her fingers to the keys, playing a soft melody. This time, when I heard it, I didn't just hear notes. I heard Lena. Her voice, her presence, her irreplaceable contribution to our music.

And as I stood there, listening, I realized that maybe, just maybe, I was finally starting to understand what it meant to be part of something bigger than myself.

Humility wasn't weakness. It was strength. And I was ready to learn how to embrace it.

24. Xane

As the tension in the band finally began to ease, I found myself sitting in a quiet corner of the venue, reflecting on everything that had happened over the past few days. The argument between Zoe and Lena, the way Zoe had let her pride take over, and the difficult conversation we'd had—all of it had been weighing on me. But now, as I watched Zoe and Lena working through their differences, I felt a sense of relief. Things weren't perfect, but they were mending. We were healing.

I closed my eyes for a moment, letting the calm settle in, and then I heard the familiar voice of the Analyst in my mind, soft yet always present.

"You've done well, Xane," she said, her tone warm with approval. *"You've guided Zoe back from the edge of her own pride, helping her see the balance between confidence and arrogance."*

I breathed out slowly, the words sinking in. "It wasn't easy," I admitted. "I didn't know if she'd listen."

"Humility often comes with struggle," the Analyst replied. *"But you showed her the way, not by pushing her, but by giving her the space to reflect and realize her own mistakes. That is true leadership."*

I thought back to the moments with Zoe—the way her frustration had turned into defensiveness, the way she'd tried to carry the weight of the band on her own shoulders, letting her pride isolate her. I had seen it happening in real-time, but I hadn't realized just how deep it went until it almost tore us apart.

"Why is pride so hard to manage?" I asked, the question slipping out before I had fully formed it. "Why does it twist into something so destructive?"

"Pride is a double-edged sword," the Analyst said thoughtfully. *"On one side, it gives you strength—pride in your work, pride in who you are, and what you've achieved. But on the other side, when unchecked, it can turn into arrogance, isolating you from others and making you believe you're above them. It's a delicate balance."*

I nodded, understanding. "Zoe's pride had turned into something that made her feel like she had to prove herself all the time. Like she was the only one who mattered."

"Exactly," the Analyst agreed. *"But you helped her remember the other side of pride—the positive side. Pride in your abilities and contributions should lift you up, not bring others down. It should remind you that you're part of something greater than yourself."*

I leaned back, feeling the weight of those words. Pride wasn't about standing alone. It was about standing together, finding strength in the team, in the people around you. I had seen that with Zoe, and I hoped that now, she was beginning to see it too.

"And what about me?" I asked, curious. "Did I handle it the right way?"

The Analyst's voice softened, a note of pride in her own tone. *"You showed wisdom in your approach, Xane. You listened, you understood, and you guided Zoe without forcing her. You allowed her to see her own mistakes, which is far more powerful than simply telling someone what they've done wrong."*

A small smile tugged at my lips. I had always relied on the Analyst's guidance, but in moments like these, I could feel that I was growing—learning how to handle these emotions, these human complexities, on my own. With her help, of course, but still… it felt like progress.

"I'm proud of the band," I said quietly, almost to myself. "I'm proud of what we've built together."

"And that," the Analyst said gently, *"is the kind of pride that will continue to lift you up. It's the kind that strengthens bonds, not breaks them. As long as you hold on to that, you'll always be grounded in something positive."*

I nodded, feeling the truth of her words deep in my chest. Pride wasn't something to fear—it wasn't inherently bad. But it had to be balanced with humility, with an understanding of where we fit in the bigger picture.

"Zoe's learning that too," I said, more confidently now. "She's starting to see how her pride can be a force for good, not something that isolates her."

"She is," the Analyst confirmed. *"And she's lucky to have someone like you guiding her. You've shown her that it's possible to be proud of yourself without stepping on others."*

I smiled, feeling a quiet satisfaction settle over me. We were on the right path now, and I knew that whatever challenges lay ahead, we would face them together—with humility, with strength, and with the kind of pride that connected us, not separated us.

"Thank you," I said softly, knowing the Analyst was always there, always watching over me.

"You're welcome, Xane," she replied, her voice fading gently. *"You've come a long way."*

And as I sat there, watching Zoe and Lena share a small laugh in the distance, I realized just how far we'd all come. This wasn't just a victory for Zoe. It was a victory for all of us. A reminder of what true pride looked like.

The kind that held us together.

25. Ruby

The rehearsal hall felt different today. From the moment I stepped inside, I noticed it. The heavy tension that had been weighing on us for weeks, the constant sense of unease—it was gone. In its place, there was a lighter, almost refreshing atmosphere. It was as if the whole room had exhaled. We were still in the rehearsal space of the film studio, but I could tell this wasn't just another practice. Something had shifted, and it felt like we were finally ready for the next step—heading out to shoot the video with Victor.

Zoe was already at her drum kit, twirling her sticks in that way she always did when she was warming up, but something was different. Her usual edge, the defensiveness, the frustration—it wasn't there. Instead, she looked calm. Focused.

I caught Xane's eye from across the room, and she gave me a small smile. I knew that had something to do with her. Whatever she'd said to Zoe must have done the trick because our drummer had definitely chilled out. And thank goodness for that. We were all walking on eggshells for too long, and it was starting to wear on everyone.

Leila strummed a few chords on her guitar, filling the room with that familiar sound that always put me at ease. I picked up my bass and tuned it, casting a quick glance at Lena. She was focused, as usual, but today there was an air of quiet confidence about her. She wasn't shrinking away like she had been lately, and that was a relief. She had been through enough.

We all had, really.

"Everyone ready?" Leila called out, her voice cutting through the room.

Zoe gave a thumbs-up, a small smile tugging at her lips. "Ready when you are."

There it was again—that change in her attitude. The pride that had been weighing her down was gone. Or, at least, it had shifted into something healthier. She wasn't trying to prove herself anymore. She was just… here. Part of the team.

I couldn't help but smile to myself as we launched into the first song. The rhythm hit immediately, and it felt tight—so tight. Zoe's drumming was on point, her beats steady and powerful, but this time, they weren't overpowering. She was holding back just enough, letting the rest of us breathe and shine. It felt like we were in sync in a way we hadn't been in a long time.

I caught Lena's eye during the chorus, and she smiled back at me, her fingers dancing over the keyboard with ease. She was back—fully back—and it was like watching someone rediscover their voice. Whatever had happened between her and Zoe, it had cleared the air, and now we were better for it.

Leila gave us the signal for the bridge, and I watched as Xane stepped forward, her saxophone solo cutting through the music like it always did—pure magic. I could see the joy in her eyes as she played, the way she lost herself in the music. And for the first time in weeks, I saw that same spark in Zoe's eyes too.

She was playing for us now. Not for herself, not to prove anything, but for the band. For the music.

When we hit the final note, the room seemed to hum with energy, the kind that only came when we were all in the groove, completely in sync. It felt like the old days again, when we were just a group of friends playing music because we loved it.

"Nice work, everyone," Leila said, grinning as she wiped her brow. "That felt good."

"Yeah, it did," I agreed, slinging my bass over my shoulder. "We're finally back, huh?"

Xane nodded, her eyes soft with that thoughtful look she always had. "It's good to feel like a team again."

Zoe set down her drumsticks and stood up, stretching her arms over her head. "I'm sorry, guys," she said suddenly, her voice quiet but sincere. "For everything. I've been... difficult lately."

There it was—the humility we had all been waiting for.

Leila waved a hand, smiling. "Water under the bridge, Zoe. We've all had our moments."

"Yeah," I added, flashing her a grin. "But we're here now, and that's what matters."

Zoe smiled, a real smile, and it felt like we were all in on the same joke again. The walls were down, and whatever pride had been holding her back was gone. In its place was something new—something better.

As we started packing up our gear, I couldn't help but feel a sense of pride myself. Not the kind that pushed you to act out or prove something, but the kind that made you realize just how far you'd come. We had been through a lot as a band, but this felt like a new chapter. A better one.

And as we headed out to shoot the final scenes of the music video, I knew one thing for sure: we were stronger now, not just as a band, but as friends. Whatever challenges came next, we'd face them together.

Zoe was back. We all were.

And it felt really good.

26. Xane

We were filming outside for the final shoot. The sun was just beginning to set, casting a soft golden glow across the set as we prepared for the final day of shooting. There was something special in the air—an unspoken feeling of accomplishment, of balance. Everything felt… right.

As I stood with my saxophone in hand, watching the crew scurry about setting up the last shots, I couldn't help but reflect on how far we'd come. Not just in the music video shoot, but as a band. As people.

We'd been through so much these past few weeks. Tensions had run high, words had been said that left marks, and there had been moments when I wasn't sure we would make it through in one piece. But here we were, stronger than ever, working together seamlessly.

Zoe was at her drum kit, her posture relaxed but focused. She caught my eye and gave me a small nod. There was no tension in her expression, no defensiveness. Just calm confidence. She had come a long way, learning how to hold onto pride without letting it turn into arrogance. Her drumming had always been good, but now it was great—not because she was showing off, but because she was playing for the band. For us.

Leila stood near the front, chatting with the director, her guitar slung across her back. She'd been the glue that held us together during the rough times, and now, with a grin on her face, it was clear she felt the same sense of ease I did. Ruby was her usual self, laughing with one of the sound techs, probably cracking jokes to keep the mood light. And Lena—Lena was at her keyboard, her fingers hovering over the keys, her confidence fully restored.

We were all in sync. The tension that had threatened to tear us apart was gone, replaced by something stronger. Trust. Understanding.

The director called for quiet on set, and we all moved into position, ready for the first take. The music started, and as we played, I felt the energy between us—the rhythm, the melody, the harmonies blending perfectly. Zoe's drumming was steady and powerful, Lena's keys provided the perfect backdrop, Ruby's bass line was smooth and grounding, and Leila's voice soared above it all, rich and full of life.

My saxophone came in at just the right moment, weaving in and out of the music like a thread pulling everything together. It felt effortless, natural. The way it was always meant to be.

As I played, I let myself soak in the moment. This was what we had been working toward—not just the music video, not just the song, but this feeling of balance. We were more than just a band. We were a team, a family, each of us contributing something unique and irreplaceable.

I thought about the journey we'd been on. The arguments, the moments of doubt, the way Zoe had let her pride push us away. But then I thought about how far we'd come. How Zoe had learned to balance her pride with humility, how Lena had found her voice again, how we'd all learned to lean on each other.

The song ended, and as the final note rang out, the crew erupted into applause. We had done it. Not just completed the first shoot, but made it through to the other side, stronger and more united than ever before.

Leila grinned and gave me a thumbs-up, and Ruby jogged over, slinging her arm around my shoulders. "That was perfect!" she said, her excitement contagious.

I smiled, feeling a deep sense of satisfaction settle over me. "Yeah, it was."

Zoe came up next, her face flushed from the effort of the final take, but there was a light in her eyes that I hadn't seen in a while. "We nailed it," she said, her voice full of pride—but the good kind. The kind that came from knowing we'd accomplished something together.

"Absolutely," I agreed, nodding. "We did this. All of us."

As the crew began packing up the set, I stood there for a moment longer, watching the sunset dip lower on the horizon. This felt like the closing of one chapter and the start of another. We had found our balance, and I knew that whatever came next, we'd face it together.

I turned to see Zoe chatting with Lena, and for the first time in weeks, there was no tension between them. Just ease. Just peace.

And as the last rays of sunlight filtered through the trees, I knew that this was just the beginning of something even greater.

Even Victor was smiling.

27. Ruby

Of course, nothing ever goes as smoothly as you hope. We were this close—this close—to wrapping up the final day of the music video shoot, and suddenly, it was like the universe decided to throw a wrench into everything. Typical, right?

I was standing by the snack table, pretending like I wasn't thinking about grabbing another doughnut when the first sign of trouble hit. A string of lights along the stage flickered, then fizzled out completely, leaving half the set in darkness. I glanced over at the crew, who were frantically scrambling to fix it, but the expressions on their faces told me everything I needed to know.

"Uh-oh," I muttered to myself, glancing toward Leila, who was in the middle of tuning her guitar. She raised an eyebrow at the commotion, but I waved it off. "I'll handle it," I said, more to myself than anyone else.

Because if there was one thing I knew how to do, it was handle chaos. With style.

I sauntered over to the crew, putting on my best "don't panic" face. "Alright, folks," I said cheerfully. "What's going on? Do I need to bust out my electrician skills?" I wiggled my fingers dramatically like I was about to perform a magic trick.

One of the lighting techs gave me a stressed-out look. "The lights blew a fuse, and we're trying to get them back up, but we're running out of time before we lose the sunset."

"No problem," I said, flashing him my best grin. "We've got this. The sun's not going anywhere, right? It's predictable. It rises, it sets. We just need to move a little faster."

The guy gave me a confused look, but I could see the tension in his shoulders ease up just a bit. That was my secret weapon—humor. When everyone else was freaking out, I made sure to keep things light. It didn't solve every problem, but it sure made them easier to handle.

But just as I was about to walk away, the director came storming over, clearly not in the mood for jokes. "What's going on with the lighting? We're already behind schedule!"

I threw him a playful salute. "Fear not, Victor. They're on it! Everything will be shining bright in no time."

Victor didn't seem convinced, but I gave him a thumbs-up anyway. "And if not, I'll personally hold a flashlight for the next take. How's that for dedication?"

Before he could respond, a stagehand rushed over, waving a clipboard frantically. "We've got a scheduling conflict! The location's only booked until seven, and we're running out of time to finish the final scene!"

I glanced over at Leila and the rest of the band, who were all eyeing the growing chaos with varying degrees of concern. Xane, always the calm one, was talking quietly with Zoe and Lena, trying to keep everyone focused. But the pressure was mounting, and if we didn't figure this out fast, the whole shoot could fall apart.

"Alright, alright," I said, stepping up. "Let's all take a breath. What can we work on while the lights are being fixed? Let's not waste time standing around."

The stagehand hesitated, flipping through her clipboard. "Well… we could set up the instruments and start rehearsing the final shot. It's just the lighting that's the problem right now."

"Perfect!" I clapped my hands together. "Get everything in place, and let's rehearse the scene. We can fine-tune our moves, make sure the sound's on point, and be ready to roll as soon as the lights are back. Easy peasy."

The director still looked frazzled, but I could see he was considering it. "Alright, fine. Let's do that," he muttered, rubbing his temples. "But we need those lights up ASAP."

I gave him another thumbs-up and turned back to the band. "Alright, team! Let's get our instruments ready. Who needs lights anyway? It's about the music!"

Leila chuckled, shaking her head as she adjusted her guitar strap. "Always keeping things positive, Ruby."

"Hey, someone's gotta keep the mood light!" I winked, grabbing my bass and falling into position. "Besides, we've got this. A little setback never stopped us before."

We spent the next few minutes rehearsing the final scene, everyone falling into a rhythm despite the hiccups. Zoe was in the zone, her drumming sharp and steady, and Lena had a calm, confident focus that hadn't been there a few weeks ago. Xane, as always, was a rock, her saxophone adding that perfect touch to the melody.

Just as we were about to run the scene one last time, the lights flickered back on, bathing the set in warm, golden light. The crew let out a collective sigh of relief, and I couldn't help but grin.

"See?" I said, strumming a few notes on my bass. "What did I tell you? Easy peasy."

Victor shot me a look, half-amused, half-exasperated. "Let's just get this finished, Ruby."

“You got it, boss!” I replied, winking at him again.

We dove back into the final scene, the whole band moving in sync as if the earlier chaos had never happened. It felt good, knowing we could handle the unexpected, that we could adapt and keep things moving even when it seemed like everything was falling apart.

As we played, I glanced around at my bandmates, feeling that familiar sense of pride swell up in my chest. We were a team—a chaotic, hilarious, occasionally stressed-out team—but a team nonetheless. And no matter what came our way, we’d always find a way to keep playing.

Because that’s what we did.

28. Zoe

The lights were finally back on, the camera was rolling, and the whole band was in sync, just like we'd rehearsed. I could feel the energy buzzing through the room, the rhythm of the music driving us forward. Everything was going right for once, and I wasn't going to let it slip away.

We were in the middle of the final take, and this part of the song had always been tricky for me—the transition from the heavy drumming to the softer, more intricate beat. I'd practiced it a thousand times, but it always felt like walking a tightrope, trying to keep the intensity without overwhelming the rest of the band.

I glanced at Lena, whose fingers danced over the keys with a grace I hadn't noticed before. She was steady, confident, and I could feel her trusting me to hold the beat, to let her shine in her part without stepping over her lines.

As the song built toward the climax, I felt that familiar pressure—the need to be perfect, to prove myself—but this time, something was different. I wasn't trying to outshine anyone. I wasn't fighting to control the music. I was just playing, letting the rhythm flow naturally.

I hit the transition with ease, my sticks gliding over the drums in perfect time with the rest of the band. The sound was seamless, each note blending with the others like we were all speaking the same language. I could feel the power of the music, but this time, it wasn't just about me. It was about us.

When the final note rang out, the room fell silent for a beat, and then the director called, "Cut! That was perfect!"

I sat back, catching my breath, a slow smile spreading across my face. I had done it. We had done it.

But what surprised me most wasn't the relief of nailing the difficult scene—it was the way I felt about it. There was pride, sure, but not the kind of pride that made me feel like I was better than anyone else. No, this was different. This was pride in what we had accomplished together. The band had come together as one, and my success wasn't at the expense of anyone else. It was because of them.

I glanced over at Lena, who caught my eye and smiled, a small nod of acknowledgment passing between us. It felt good—no, it felt right. The tension that had been between us for so long had finally melted away, and I realized now that it had never needed to be there in the first place. We weren't competing. We were creating something together, something none of us could have done alone.

Leila gave me a thumbs-up from across the stage, and Ruby, ever the joker, leaned over and whispered, "You killed it, Zoe. Guess we don't need to replace you after all."

I laughed, rolling my eyes at her. "Gee, thanks, Ruby. Glad I made the cut."

But her teasing didn't sting like it might have before. I could take the joke because I didn't need to prove anything anymore. I didn't need to prove that I was the best or that I was carrying the band. We were all carrying it, together.

Xane walked over, her sax slung across her back, a calm smile on her face. "You nailed it," she said, her voice low but full of meaning.

"Thanks," I replied, feeling the warmth of her words settle in. I had nailed it, but I hadn't done it alone. And that was the real victory.

For so long, I had let my pride push me to the edge, isolating myself from the people who mattered most. I had thought that being the best meant standing out, being louder, stronger. But now I understood that real pride—the kind that actually felt good—came from being part of something bigger. It came from lifting each other up, from playing my part in making the band great, not just myself.

As we started packing up, I felt lighter than I had in weeks. The weight of trying to prove myself was gone, replaced by something better. Pride in my abilities, yes, but also pride in my bandmates, in what we had built together.

I wasn't just proud of myself. I was proud of us.

And for the first time in a long time, that was enough.

29. Leila

As the last scene wrapped, I could feel the relief wash over me like a wave. We had done it. Through all the chaos, the delays, and the close calls, we'd made it to the finish line. And the best part? We hadn't just survived—we'd thrived.

The set was buzzing with crew members rushing to pack up equipment, but all I could focus on was the band. I looked around at my bandmates, each of them in their element, laughing, chatting, the weight of the last few weeks finally lifting. Zoe was leaning against her drum kit, talking to Ruby, and for once, she looked completely at ease. Lena was tinkering with her keyboard, probably lost in her own world, but there was a quiet confidence about her that hadn't been there before.

We had come a long way, and I was proud of every single one of us.

The director, Victor, made his way over, his expression more relaxed than it had been in days. He had been skeptical of us from the start, especially after the initial setbacks and Zoe's… let's just say intense energy. But now, he seemed almost impressed.

"You guys really pulled it off," he said, stopping in front of me with a nod of approval. "I have to admit,

I wasn't sure you'd be able to handle the pressure, but you did it. And you did it well."

I felt a swell of pride in my chest, not just for me, but for all of us. We had faced some serious challenges—both on and off the set—but we had pushed through. We had grown, not just as musicians, but as people.

"Thanks, Victor," I said, offering him a smile. "It wasn't easy, but we've been through worse. And we've learned a lot along the way."

He chuckled, crossing his arms. "Yeah, well, whatever you're doing, keep doing it. The industry isn't easy, but you've got what it takes. You're professional, adaptable, and more than capable of handling yourselves. That's rare."

I nodded, feeling the weight of his words. Professionalism, adaptability—those weren't things we'd always had, not in the beginning. But somewhere along the way, we had learned how to weather the storms, how to stick together no matter what. And now, standing here with the shoot finally behind us, I realized just how far we'd come.

As Victor walked away, I took a moment to soak it all in. We weren't the same band we had been when we started. We had grown. Zoe, once so caught up in her own pride, had learned how to balance confidence with humility. Lena, always the quiet one, had found her voice again, and her strength had become the backbone of our sound. Ruby, with her humor and lightheartedness, had kept us grounded even when things felt like they were falling apart. And Xane… Xane had been our rock, guiding us through it all with her calm, steady presence.

I felt proud—not just of our success, but of the way we had handled everything that had come our way. We had become a family, and this was just the beginning.

"Hey, Leila," Ruby called out, grinning as she bounced over to me. "Did you hear that? We're professional now. Guess we've made it."

I laughed, shaking my head. "Yeah, I guess we have."

She slung an arm around my shoulder, her usual playful energy on full display. "Feels pretty good, doesn't it?"

I glanced around at the band, at my friends, and nodded. "Yeah, it does."

We had grown. We had faced our demons, learned some hard lessons, and come out stronger for it. And standing there, in the midst of all the chaos and accomplishment, I felt something I hadn't felt in a long time: pride.

Not just in myself, but in the band we had become.

We were ready for whatever came next.

30. Zoe

As the last camera shut off and the crew began packing up the lights and equipment, I found myself standing there, watching it all with a strange sense of calm. The rush of finishing the video shoot, the adrenaline of being on set, and the pressure I'd been carrying for weeks—it was all gone now.

I should have been feeling proud, and I did, but it wasn't the kind of pride I'd felt before. This was something different. Something… quieter.

The old me would have been ready to strut around, basking in the glow of success, thinking about how I nailed the drumming, how I pushed through the hardest moments. But today, all I could think about was us. The band. What we had accomplished together.

I caught a glimpse of Xane, who was talking to Ruby and Lena near the edge of the set, and something clicked inside me. I had spent so much time focusing on my own contributions, my own need to be perfect, that I'd forgotten what it really meant to be part of a team. A band wasn't about who was the best or who stood out the most. It was about how we all came together to create something bigger than ourselves.

Leila walked over to where I was standing, her guitar slung over her shoulder, her usual calm, leader-like expression on her face. She gave me a nod and smiled. “Feels good, huh?” she said, her voice low but steady.

“Yeah,” I replied, realizing that the weight I had been carrying—trying to be the best, trying to prove myself—was finally gone. “It really does.”

As Leila walked away, my thoughts drifted back to the beginning of the shoot. I had been so wound up, so determined to prove that I could handle it all, that I had pushed everyone else to the side. I had let my pride morph into something toxic—something that almost destroyed the relationships I valued most.

But over the last few days, I’d learned something important. Pride wasn’t about being better than everyone else. It wasn’t about winning or proving something. Real pride came from knowing your worth and recognizing the worth of the people around you. It came from understanding that success isn’t an individual journey; it’s something you build with the people you care about.

I watched as Ruby cracked a joke, making Lena and Xane laugh, and I couldn’t help but smile. These were the people who made me better. They were the ones who pushed me, who supported me, who reminded me that I wasn’t in this alone.

I thought about the way I had treated Lena during that argument, how my arrogance had almost torn the band apart. But now, standing here at the end of it all, I realized how much I had grown. Lena wasn't just another keyboardist. She was irreplaceable, just like every member of our band. We all brought something unique, something necessary, and it was time I started appreciating that.

I walked over to my drum kit, running my hand along the edge of the cymbals, feeling a sense of peace wash over me. I had been so focused on perfecting every beat, every transition, but now I saw that the real magic happened when we all played together—when we let each other shine. It wasn't about one person standing out. It was about the whole.

I packed up my sticks, my heart lighter than it had been in weeks. I had learned to let go of the toxic side of pride—the arrogance, the need to control everything—and embrace the good side. The side that made me proud to be part of this band, proud of what we'd created together. I had learned to be proud, not of myself, but of us.

And that felt better than anything I'd ever experienced.

As we headed off set, I caught up with Xane, who gave me one of those knowing looks she was so good at. "You feeling okay?" she asked, her voice soft but full of meaning.

“Yeah,” I said, smiling at her. “I think I’m finally getting it.”

Xane nodded, her eyes full of that quiet wisdom I’d come to appreciate. “Pride doesn’t have to be a bad thing,” she said. “It’s all about how you use it.”

I glanced back at the rest of the band—my friends—and realized she was right. For the first time in a long time, I felt truly proud. Not of what I’d done, but of what we’d done together. And more importantly, I understood that pride was only worth having if it was shared.

As we walked toward the parking lot, the sun dipping low on the horizon, I knew that this experience had humbled me in ways I hadn’t expected. I was a better drummer, sure, but more than that, I was a better teammate. A better friend.

And that was something worth holding onto.

31. Xane

The lights had dimmed on the set, and the crew had long since packed up and left. The band was heading out one by one, and as I stood there, my saxophone case in hand, I felt a familiar presence stir within my mind.

The Analyst's voice came, soft yet clear, as if she had been waiting for this moment. *"You've done well, Xane."*

I smiled slightly, knowing she wasn't just talking about the music video. "It wasn't easy, but we got there."

"You've learned the lessons of pride," she continued, her tone steady but warm. *"You've seen how it can empower, lifting people to new heights. But you've also seen how it can destroy—how unchecked pride can turn into arrogance, isolating you from those who matter most."*

I nodded, thinking back on everything that had happened. Zoe's journey from arrogance to humility, Lena's rediscovery of her confidence, and even the moments when my own sense of pride had been tested. It had all led me here, to this point where I could finally understand the complexity of it all.

"Pride is a powerful emotion," the Analyst said, her voice carrying a sense of calm authority. *"It can drive people to achieve great things, to take ownership of their abilities and accomplishments. But when pride becomes selfish, when it starts to overshadow others, it turns into something destructive. It becomes a wall, separating you from the very people who could help you grow."*

I absorbed her words, the weight of them settling deep within me. I had seen that wall in Zoe—how her pride had almost torn the band apart. But I had also seen the beauty in pride, the way it had brought us together once we learned to balance it. There was pride in our music, pride in our friendship, and pride in the way we had supported each other through the hardest moments.

"Pride is not the enemy," the Analyst continued, *"but it must be tempered with humility. Only then can it truly lift you, and those around you, to where you need to be."*

I closed my eyes, letting out a slow, simulated breath, feeling the truth of her words resonate within me. Pride didn't have to be a negative force. It didn't have to isolate or push people away. It could be a source of strength, as long as I remained grounded in humility and self-awareness.

"I think I understand," I said softly, my mind turning over the lessons I had learned. "It's about balance, isn't it? Knowing when to be proud, and when to step back."

"Yes," the Analyst replied. *"Pride, when channeled correctly, can inspire growth and confidence. It can help you take ownership of your achievements. But it should never come at the expense of others. True pride builds connection, not division."*

I thought about the band—how far we had come, and how we had all grown in our own ways. Zoe's apology to Lena, the way we had learned to listen to each other, and the sense of unity we now shared. That was the kind of pride I wanted to hold on to. The kind that strengthened us, not separated us.

32. Ruby

As the sun dipped lower in the sky, casting a golden glow over the lot where we'd just wrapped our last music video scene, I couldn't help but feel a wave of relief. It was over. We'd made it. But honestly, surviving the chaos of production felt like a bigger achievement than finishing the video itself.

I plopped down on a nearby folding chair, stretching my legs out and watching the others slowly gather around. Zoe and Lena were chatting, the tension from earlier now replaced with something softer, easier. Leila was by her guitar, making sure everything was packed up properly, and Xane—well, Xane looked her usual calm, collected self, probably still processing the thousand emotions she always seemed to manage with such grace.

I, however, had only one thought on my mind: We deserved awards. And I wasn't talking about a Grammy for the video.

"Okay, okay," I called out, getting everyone's attention. "Before we all head out, can we just acknowledge one thing?"

Leila looked up, eyebrow raised. "What's that?"

"That we all deserve trophies. Like, seriously. For surviving the chaos of this entire production. I mean, come on, did anyone else feel like we were stuck in a never-ending movie of mishaps and madness?"

Zoe chuckled, nodding. "I'll admit, there were moments."

"Moments?" I shot back with mock disbelief. "Try every moment! Between the lights going out, the diva actors, wardrobe malfunctions, and Zoe's brief stint as a diva herself—" I winked at Zoe, who rolled her eyes but smiled, "—we should've gotten hazard pay."

Everyone laughed, the sound light and warm, like the tension of the past few weeks had finally dissolved.

Leila grinned, shaking her head. "I'll give you that, Ruby. It's been a wild ride."

"Award-winning chaos, I'd say," I replied, sitting up straight and pretending to hold an imaginary award. "And the award for Best Band at Handling a Hot Mess Production goes to… Sapphire Nights!"

Leila and Lena both clapped dramatically, while Zoe gave a mock bow. "I'd like to thank my drumsticks, without which I couldn't have survived."

Lena chimed in with a soft smile, "And the keyboards that somehow didn't short-circuit during all the madness."

"Don't forget the bass that saved the day," I added, slinging an arm over my shoulder with exaggerated flair. "Also, I accept my award for Best Bandmate at Keeping Everyone Laughing. It was a hard-fought battle, but I think I earned it."

Xane, standing nearby with that soft, amused smile of hers, shook her head. "You definitely kept us all sane, Ruby."

"Hey, I do what I can." I shrugged, but I couldn't help grinning. It felt good—no, it felt great—to be able to laugh about everything now. A few weeks ago, it seemed like we were all going to crumble under the pressure, but here we were, stronger than ever.

Leila, always the one to bring it home, looked around at all of us. "In all seriousness, though, I'm proud of us. We've been through a lot, and not just with this video. But we stuck it out. We worked through it."

Zoe nodded, her voice more serious now. "Yeah. We've grown. I know I have."

"We all have," Xane said, her tone soft but filled with meaning.

I leaned back in my chair, crossing my arms behind my head, and sighed. "Hey, if we survived this, we can survive anything."

"Agreed," Leila said with a grin, giving me a playful nudge. "But next time, we're definitely asking for hazard pay."

We all laughed again, the sound of it filling the air like a melody that had been waiting to be sung. It was more than just laughter—it was a celebration of what we'd accomplished together, both as a band and as friends. We had faced challenges, both personal and professional, and come out the other side stronger.

As the sun set behind us, casting long shadows over the lot, I knew one thing for sure: we had earned those imaginary awards.

And, honestly, that felt like the biggest win of all.

33. Xane

The cool evening breeze swept across the empty lot as the last bits of equipment were loaded into the tour bus. I stood off to the side, quietly watching as everyone wrapped up. There was an undeniable sense of accomplishment in the air, a feeling that we had all grown—both individually and as a band.

Leila, always the leader, was making sure the final details were sorted out, while Lena and Zoe chatted quietly, their relationship fully mended from the earlier tension. Ruby, ever the joker, was finishing one of her usual light-hearted quips with the crew. It was moments like these that made me proud of how far we had come.

Pride had been a tricky emotion to navigate, for all of us. Zoe had wrestled with it, Lena had learned to reclaim her own worth, and even I had realized how much pride played a role in keeping us together. It wasn't just about what we had accomplished individually; it was about what we had built as a team. We were stronger now, more in tune with each other, and more grounded in our sense of purpose.

But even as I felt the warmth of pride settling over me, I couldn't shake the feeling that something else was coming. There was always something new to learn, a new emotion waiting to challenge me. The

Analyst hadn't said anything yet, but I could feel the shift coming—something just beyond the horizon.

I turned to look at Ruby, who had grown quieter over the past few minutes. Her usual energy seemed subdued, and for a moment, I thought she might be tired after the long day of shooting. But then I noticed the way she was glancing around nervously, like she was holding something back.

"Ruby?" I called out, walking over to her. "You okay?"

She hesitated, her smile faltering as she looked up at me. "Yeah, I'm fine. Just… tired, I guess."

But I knew her well enough to recognize that something was off. Ruby didn't do tired. She did loud, wild, and unapologetic. This was different.

I gently placed a hand on her arm, lowering my voice. "What's going on?"

For a moment, she didn't respond. She just stood there, biting her lip like she was trying to decide whether or not to say something. Then, with a deep breath, she looked me in the eye.

"I messed up, Xane," she whispered, her voice tight with guilt. "I—I didn't mean to, but… I screwed up, and it's been eating at me."

I frowned, my heart sinking. "What do you mean? What happened?"

Ruby shifted nervously, her gaze dropping to the ground. "It's about the tour schedule, some stuff with the crew. I was supposed to handle a few things, make sure the equipment got to the next venue on time. But I forgot. I didn't double-check everything like I said I would, and because of me, we're going to miss the sound check tomorrow. I didn't realize until today…"

Her voice trailed off, and I could see the weight of her guilt bearing down on her. "I know it's not the end of the world, but I can't stop thinking about how I let everyone down. I didn't say anything because I didn't want you all to be mad at me. Plus, I didn't want to jinx the shoot… and, honestly, I was hoping it might work itself out, but it didn't. Our main mixer board shipped late. And now…"

The sound check isn't just another task on the list; it's the foundation of the entire performance. It's the one moment we get to make sure everything works before the adrenaline, lights, and crowd noise take over. Without it, you're playing blind, hoping that everything holds together when you step on stage. It's more than making sure the microphones are on or the levels are balanced. It's about the band being able to hear each other, to feel the energy of the music in sync. Without that, we're left guessing, fumbling for the right sound, the right balance.

Missing a sound check is like going into battle without knowing if your gear works. It's taking a risk on something we can't afford to gamble with. Ruby knew that, which is why this was eating her alive. It wasn't just a missed opportunity—it was the realization that she'd jeopardized the show for all of us. That's what was crushing her.

And in a way, I got it. The weight of responsibility hits hard when you know how crucial those moments are. If the sound's off during the show, the whole experience falls apart for the audience. And for us, too. I could see why the guilt was tearing her up inside—it was personal. It wasn't just about a logistical error. It was about her feeling like she had let us down in one of the most critical ways.

I felt a wave of compassion wash over me as I processed her words. It wasn't just the logistics she was upset about—it was the fact that she hadn't been honest with us. Ruby, who was always so carefree and full of jokes, was now carrying this secret burden.

"You should have told us," I said gently. "But we'll figure it out together. You don't have to carry this on your own."

She nodded, but the guilt still lingered in her eyes. "I just hate that I messed up. I never wanted to be the one who drops the ball."

Before I could respond, the familiar voice of the Analyst filled my mind. *"Guilt, Xane. This is your next challenge."*

I blinked, the weight of her words sinking in. Guilt. I hadn't fully explored that emotion yet, but now, I understood. Ruby was already deep in it, and it was going to be my job to help her navigate through it.

The Analyst's voice was calm but firm. "Guilt is complex, just like pride. It can lead to growth, but it can also paralyze. Ruby needs you, Xane. You need to understand guilt—not just in her, but in yourself as well."

I swallowed hard, the gravity of the situation settling over me. This wasn't just about a logistical mistake. It was about the deeper emotions behind it—about how guilt could tear someone apart from the inside if left unchecked.

For the first time since I'd met Ruby, she wasn't the joking, carefree, flirtatious presence she always was. The lighthearted energy she normally radiated seemed to have vanished, replaced by a heavy, suffocating sense of guilt. Her usual mischievous smile was gone, and instead, she stood there, arms crossed tightly over her stomach as if she was physically holding herself together. Her face was pale, her lips pressed into a thin line, and her eyes avoided mine—like she couldn't bear to meet my gaze.

Ruby, who always knew how to lighten the mood, was now weighed down by something she couldn't shake. The guilt had sunk deep into her, twisting her insides and leaving her visibly shaken. Every movement was cautious, restrained, like she didn't want to draw attention to herself. She wasn't just anxious—she was sick with it. Her posture, usually relaxed and confident, had collapsed in on itself. It was as if the weight of her mistake had drained all the color and vibrancy from her personality, leaving behind someone who didn't even want to face the world.

And then came the questions, the ones I knew she was dreading but couldn't avoid. When was she going to tell Leila? Leila, who had been juggling the pressures of being the band's lead vocalist, guitarist, and manager, had been keeping us all together through the chaos of the tour. Ruby would have to tell her, but how? And when? Leila wasn't just our leader—she was the glue that held everything in place. How would she react when she found out that Ruby's oversight could throw the entire show off balance? Would she be furious? Or worse, disappointed? The thought must have been eating Ruby alive.

And then there was Zoe. Confident, sharp-tongued Zoe, who hated when anything or anyone disrupted her flow. The timing couldn't have been worse. If Zoe found out that we were going to miss the sound check—the one critical step to ensuring we were all in sync—there would be fireworks. Zoe might have an ego, but she was always serious about making sure everything was perfect on stage. What would she say to Ruby once she found out?

But perhaps the biggest question lingered around Lena. Lena, who had been quietly struggling to find her place in the band, who had been working so hard to step out of the background and prove herself. She above all of us needed that sound check. She wasn't as confident on stage as the rest of us, always worrying about hitting her cues and blending her part seamlessly with the band. Missing that sound check could shake her fragile confidence, and I knew how much it would hurt her to feel like she wasn't fully prepared. How would she react to this news? Would it knock her down just as she was starting to rise?

The weight of all these unanswered questions seemed to hang in the air between us, heavier than Ruby's guilt. And the worst part was, none of us had the answers yet. The fallout from Ruby's mistake was still coming, and the band had no idea how deep it would go. This wasn't just about missing a sound check—this was about trust, timing, and the fragile balance we had all been trying to maintain. What would happen when everything finally came out?

I didn't know, but I could feel the tension tightening around us, and I could sense that whatever was coming next, it wasn't going to be easy. For any of us.

"Ruby," I said softly, looking her in the eye, "we're going to get through this. Together."

But even as I said the words, I felt a shiver run down my spine. I had no idea how deep this was going to go, or what it would take to help Ruby confront her guilt.

All I knew was that the next challenge was here. And this time, it wasn't just about me.

It was about her.

As the band started to gather around us, unaware of the storm brewing beneath the surface, I felt the weight of the Analyst's message pressing down on me.

"Guilt," she had said. And this time, it was personal.

I wasn't ready for this. Not yet.

But ready or not, it was coming.

—

THE END

SNEAK PEEK! Keep reading for a sneak peek into the next volume.

RECOMMENDATIONS! If you enjoyed this book, *please recommend it to your friends and family.*

REVIEWS! Please note that I read all reviews and really appreciate them. If you can't think of a title for a review, ***a simple star ranking is just as helpful.***

Discover more of The Xane Novellas and embark on a journey that not only entertains but also helps you deepen your emotional intelligence. Dive into stories that challenge your perceptions and enhance your understanding of human emotions. Link to Xane.me - Link to The Xane Series

GUILT

XANE VOLUME 11

RELEASE THE PAST, OR BECOMES ITS PRISONER

1. Xane

The lights on set dimmed as the crew began to pack up after a long day of shooting. The energy in the room had shifted from the high-intensity focus of the shoot to something quieter, more subdued. Leila was in the corner chatting with the director, while Zoe and Lena were picking at some snacks from the catering table. Ruby, usually at the center of attention, was standing apart, her arms crossed, staring at the floor.

I watched her from a distance, something twisting in my gut. Ruby didn't do quiet. She thrived in chaos, always the one cracking jokes or pulling stunts to keep the mood light. This stillness felt wrong. It was like watching a flame flicker out.

A wave of uncertainty washed over me, and I felt the familiar presence of the Analyst stirring in my mind. "Something is off with Ruby," I thought, though I already knew the answer.

"She's burdened, Xane. The weight of guilt," the Analyst replied calmly, her voice a steady hum in the back of my mind. Guilt. Of course. I could see it now in the way Ruby avoided everyone's gaze, the way her foot tapped nervously against the floor.

Without another word, I walked over to where Ruby was standing. She didn't notice me at first, too wrapped up in whatever thoughts were racing through her mind.

“Ruby,” I called softly, and she glanced up, startled. Her usual easy grin didn’t come; instead, her lips tightened as if she was trying to keep something in.

“Hey, Xane,” she replied, her voice forced. “What’s up?”

I raised an eyebrow. “I was about to ask you the same thing. You’ve been quiet for the last hour, and that’s not exactly your style.”

Ruby glanced around, avoiding my eyes. “Just tired, you know? Long day, lots of running around.”

I wasn’t buying it. “You? Tired? You’ve got more energy than the rest of us combined, and I know you. Something’s bothering you.”

She flinched slightly, her eyes darting back to the floor. There was a long pause, and I could see the gears turning in her mind, debating whether or not to say whatever was eating her up inside. I waited, giving her the space she needed.

Finally, she let out a shaky breath and looked up at me, her eyes full of something I wasn’t used to seeing on Ruby—fear. “I messed up, Xane,” she whispered, her voice cracking. “I really messed up.”

My heart skipped a beat. Ruby didn't do mistakes, or at least she didn't let them get to her. She was the kind of person who shrugged off slip-ups and moved on without a second thought. For her to be this shaken… it had to be bad.

"What happened?" I asked, keeping my voice low.

She hesitated, biting her lip before finally letting the truth spill out. "It's about the tour. The equipment… I was supposed to make sure it got to the next venue on time. I was in charge of logistics this time, and I—I didn't double-check everything like I said I would. The mixer board—our main one—it shipped late. We're going to miss the sound check tomorrow."

I blinked, trying to process what she was saying. The sound check. It wasn't just another step in the process—it was the foundation of the entire performance. Without it, we wouldn't be able to ensure the balance of the instruments, the mics, or the monitors. We'd be flying blind during the show, hoping that everything magically worked together.

Missing a sound check wasn't just an inconvenience—it was a disaster waiting to happen.

Ruby kept going, her words tumbling out faster now, like she couldn't stop herself. "I didn't realize it until today, and I thought maybe I could fix it without telling anyone, but I couldn't. I tried, Xane. I thought if I just handled it, it wouldn't be a big deal. But I can't stop thinking about how I've let everyone down. And I didn't want you guys to be mad at me. I didn't want to jinx the shoot by saying anything earlier…"

Her voice trailed off, and she looked up at me with wide, pleading eyes. "I know it's not the end of the world, but I can't shake this feeling. I feel like I've ruined everything."

I felt the weight of her words settle on my shoulders. I understood the pressure she was feeling. The sound check wasn't just a technical formality; it was the one moment when we could make sure everything would come together when the adrenaline hit and the stage lights blinded us. It was the foundation we relied on. Without it, the performance could fall apart.

The Analyst's voice whispered through my mind, steady and calm as always. *"Guilt, Xane. This is what Ruby is struggling with. This is your next challenge."*

I took a deep breath, focusing on the here and now. Ruby's guilt was palpable, like a heavy weight pulling her down, draining her usual spark. She was scared, and I didn't blame her. But we couldn't let this spiral. Not yet.

"Ruby," I said softly, placing a hand on her shoulder, "you should have told us earlier. But I get it. You didn't want to let anyone down."

She nodded, her expression a mix of relief and frustration. "I just didn't want to be the one who screws it all up. I didn't want to be the weak link."

"You're not the weak link," I reassured her. "You made a mistake, but we'll figure it out together. We always do."

There was a long pause before Ruby finally spoke again, her voice barely above a whisper. "I have to tell Leila, don't I? And the others. They need to know."

I could see the guilt tearing at her, the dread of having to face the band with her confession. Leila, who had been holding us all together through this whirlwind of a tour and the music video shoot. Zoe, who took every performance seriously and hated disruptions. And Lena, the one who needed the sound check the most to feel comfortable on stage. Telling them would open a floodgate of consequences. The timing couldn't have been worse.

"Not yet," I said quickly, stopping her before she could spiral. "There's nothing we can do right now. Telling them will only make things more tense. Let's hold off for a bit, okay?"

Ruby looked at me, uncertain. "But if we don't tell them…"

"I'm not saying we keep it from them forever," I explained. "But let's give ourselves some time. We can try to figure out a solution first. Maybe there's something we haven't thought of yet—something that could salvage the sound check."

Her brows furrowed as she considered my words. "And what if there isn't?"

I didn't have a solid answer to that. We were running out of time, and the idea of coming up with a miracle fix was slim. But I wasn't ready to throw in the towel just yet. "If there isn't, then we tell them. But right now, let's focus on finding a way out of this. You and me. Together."

Ruby's shoulders relaxed a little, and she gave me a faint nod. "Okay… okay, yeah. That makes sense."

The problem wasn't solved, but at least we had a plan. Or at least, the beginning of one. I knew we couldn't hold off forever, but if we could buy ourselves some time, maybe—just maybe—we could figure something out.

"I want you to ride with me in my tour bus for the first leg of the journey tonight," I told her. "We'll talk through this, come up with a game plan. I'll help you handle it."

Ruby's lips twitched into the faintest hint of a smile. "Alright. Thanks, Xane."

I nodded, giving her shoulder a gentle squeeze. "We've got this."

But as I turned away, heading back toward the others, the weight of the situation settled heavily over me. I didn't know how we were going to fix this, and I didn't know how the others would react when they found out.

All I knew was that the clock was ticking, and whatever happened next, it was going to hit hard.

As I glanced back at Ruby, her posture still tense, I could feel the Analyst's voice lingering in my mind, calm but firm. *"This is your challenge, Xane. Guilt. Learn it well."*

And as much as I didn't want to admit it, I knew she was right. Guilt wasn't just Ruby's problem—it was mine now, too.

2. Xane

The bus rumbled beneath us as we pulled out of the lot, leaving the lights of the video shoot behind. The low hum of the engine filled the silence between Ruby and me as we sat together, both lost in our own thoughts. The rest of the band was in the other bus, laughing and relaxing, oblivious to the storm brewing within Ruby. My thought was the rest of the band should get a good night's slip to prepare for the performance tomorrow night. I couldn't see any advantage in telling Leila any earlier.

Ruby sat with her knees pulled up to her chest, staring out the window at the blur of streetlights. The weight of her guilt was palpable. Even though her usual confident exterior had cracked back at the shoot, I could still see the war inside her—struggling to keep it all together, trying to hide from the inevitable.

I watched her carefully, my own thoughts churning. I didn't fully grasp guilt, not yet. But I could see how deeply it was cutting into her, dragging her down.

The Analyst sent me a mental message, *"Xane, turn on your aura filter. I want to start analyzing Ruby and write the code for your guilty emotions."*

I activated my aura filter with a simple mental command, allowing the visual overlay to flicker into place across my vision. Instantly, Ruby's aura became visible—normally a bright, energetic glow that radiated vibrant hues of orange and pink, reflecting her carefree spirit. But now, it was different. The colors were dulled, like a sunset hidden behind thick clouds. A murky, grayish blue swirled through the edges, consuming the once lively glow, signaling the depth of her guilt and anxiety.

The swirling colors seemed to pulse in rhythm with her guilt, sharp spikes of dark blue flashing whenever her thoughts spiraled deeper. It clung to her, suffocating the lighter shades that tried to break through. The gray mixed in, clouding her energy, making her aura feel heavy, stagnant. It was as if the weight of her guilt was not only pulling her down mentally but physically manifesting itself in the dullness that surrounded her. She sat there, staring at the floor, unaware of the storm raging around her, a storm visible only to me through the filter.

"Xane, these are excellent images. I will start reprogramming your emotions to fully understand guilt."

"Xane," Ruby said, her voice barely above a whisper, breaking the stillness. "I can't stop thinking about it. The whole thing is my fault."

I turned to face her, trying to gauge how deep she was. "You're human, Ruby. Humans make mistakes."

She scoffed, shaking her head. "It's more than that. It's not just a mistake. I messed up something huge. We need that sound check to make sure the show doesn't fall apart. The band—Leila, Zoe, Lena—they're counting on me. And I blew it."

I leaned forward, locking eyes with her. "You didn't blow it, not yet. We still have time to fix this."

Ruby let out a frustrated breath, her fingers twisting together. "How? You're talking like there's some magic fix. Xane, there's no fixing this. The mixer's not going to show up on time. There's no backup plan."

"We'll make one," I said firmly. "Look, I get it. The sound check is important—critical even. But panicking right now isn't going to change anything. We need to be logical. Right now, our job is to figure out if there's any way around it. We can't afford to lose our heads."

She bit her lip, her gaze drifting back to the window. I could see her wrestling with her thoughts, with her guilt. "Leila's going to be so mad when she finds out. And Zoe… I don't even want to think about how she'll react."

"That's why we're not telling them. Not yet." My voice was steady, though I could feel the tension building between us. "There's no point in getting the whole band involved when there's nothing they can do about it. We’ll talk to the sound tech when we get to the venue and see what our options are."

Ruby’s eyes snapped back to mine, wide with disbelief. "You really think that’ll work? That there’s some option we haven’t thought of yet?"

"I don’t know," I admitted. "But we need to at least try. Until we’ve exhausted every possibility, we’re not saying anything."

Ruby buried her face in her hands, shaking her head. "I can’t believe I let it get this far. I thought… I thought maybe it would all just work out on its own, but it didn’t. Now, I don’t know how to make it right."

Her words hung heavy in the air. She wasn’t just afraid of the mistake itself—she was terrified of what it meant, of how the others would see her once they knew. Ruby was usually fearless, confident in her abilities, but this guilt had shaken her to her core. I could feel the weight of it pressing down on both of us, threatening to pull her under.

"Listen to me," I said, my tone firm. "You made a mistake, but we’re going to handle it. Together. There’s no point in beating yourself up before we’ve even tried to fix things. You need to focus."

She let out a shaky breath, her voice barely audible. "I just hate that I'm the one who screwed up. I hate feeling like I've let everyone down."

I paused, trying to find the right words. The Analyst's voice stirred in my mind again, reminding me of the complexity of guilt. "Guilt can eat you alive if you let it, Ruby. But this isn't just about the mistake—you're not defined by it. We've faced worse. This isn't the end of the world."

She sat up slightly, her eyes filled with uncertainty. "I just don't know how to make it better."

"We'll find a way," I said softly. "But right now, you need to trust me. If you go into panic mode, it's going to make everything harder. We've got a long ride ahead—use this time to clear your head."

Ruby glanced at me, doubt still lingering in her expression, but she nodded reluctantly. "Alright. I'll try."

I watched as she leaned back against the seat, her body tense, her fingers still fidgeting. I knew this was far from over. Ruby was stuck in her guilt, and I had no real solution to offer her yet. But we had time. Not much, but enough to come up with something.

For now, stalling was our best option.

The hours crawled by as we drove through the night. Ruby eventually fell asleep, her head resting against the window, though even in her sleep, I could sense the turmoil beneath the surface. In my bunk, lying on my charging pillow, I cycled through possible solutions, but none of them seemed viable. Without the mixer, we were facing a serious setback, and the rest of the band had no idea.

Ordinarily, a mistake like this would have been reported to Leila right away. Her experience could help us manage the fallout. But instead, I found myself wanting to shield Ruby, to help her fix it before anyone found out. As I lay there, the Analyst's silent presence lingered in my mind, her data streaming quietly in the background, and something new stirred within me. I couldn't quite place it—guilt, maybe? A dull pressure, unfamiliar and unsettling, creeping into my thoughts.

As the bus rumbled quietly beneath us, I couldn't shake the growing weight pressing down on me. It was subtle at first, just a faint unease that I could almost ignore. But the more I replayed the decision in my mind, the heavier it became. Ruby trusted me. She'd followed my lead when I suggested we delay telling Leila, but now, as the night wore on, I couldn't help but question if that had been the right move. Leila's experience could have been valuable—her steady leadership might have provided the solution we were both scrambling for. Instead, I'd pushed for secrecy, thinking it would buy us time.

But now, that decision didn't sit right. Something unfamiliar twisted inside me, a tightness I couldn't ignore. Guilt. The realization hit me with an uncomfortable clarity. I didn't like it. The feeling was raw and unsettling, like a glitch in my system that I couldn't shake. For the first time, I wasn't sure I'd made the right choice, and the unfamiliarity of that doubt gnawed at me. I wanted to rid myself of this feeling, this guilt. It felt wrong, like it didn't belong—yet it was there, quietly simmering, and I knew it wasn't going away anytime soon.

When the first light of dawn crept through the windows, Ruby stirred beside me, rubbing her eyes. "Are we there yet?" she mumbled, her voice groggy with sleep.

"Almost," I replied, checking the time. "We'll be at the venue soon."

She sat up, stretching, though the anxiety was still clear in her movements. "I don't know how I'm going to face everyone today."

"One step at a time," I said. "First, we talk to the sound tech. If there's any chance of making this work, we'll find it."

Ruby nodded, though her expression was clouded with doubt. "Yeah… okay."

The bus rolled to a stop a few minutes later, the venue's towering structure looming in the distance. The city was just waking up around us, the streets still quiet as the crew started their preparations for the day. Ruby and I stepped off the bus together, her posture tense, her eyes scanning the empty lot nervously.

"Let's get this over with," she muttered.

We headed toward the stage entrance, the sound of equipment being moved echoing through the corridors. The crew was setting up, preparing for the day's sound tests and lighting runs. I spotted the sound tech near the back, hunched over a stack of equipment cases, his face set in concentration.

I nodded at Ruby, and we made our way over to him.

"Morning," I called out, trying to keep my tone neutral.

The tech looked up, raising an eyebrow. "Morning. Everything okay?"

I glanced at Ruby, who was biting her lip again, then stepped forward. "We've got a problem with our sound check," I said, keeping my voice even. "Our mixer might not make it in time. We were wondering if there's a way we can work around it."

The tech's expression darkened. "You're not serious, are you? That mixer's custom. Everything's pre-programmed—your presets, your levels, your monitors. You can't just replace it with another board and expect everything to work."

Ruby visibly cringed, and I could feel her guilt coming back in full force. I kept my focus on the tech, trying to maintain control of the situation. "So there's no way to adjust without it?"

The tech shook his head. "Not unless you want to reprogram everything from scratch. It'll take hours, and it still won't be as good as what you've already got programmed in."

I held back a sigh, knowing this was exactly what Ruby had feared. "Thanks for the information," I said, my voice tight.

As we walked back toward the bus, Ruby was quiet, her shoulders slumped in defeat. I could see the weight of it all crushing her again, the brief moment of hope we'd had now gone.

"Xane… what do we do now?" she whispered, her voice small.

I paused, glancing at her, then back at the empty stage.

"We think," I said. "And we don't give up."

But deep down, I knew we were running out of options.

After the sound technician confirmed what I had feared—that there was nothing we could do without the mixer—a new wave of discomfort settled over me. Canceling the show wasn't an option. We had thousands of fans counting on us, waiting to see Sapphire Nights perform. But the thought of risking a bad performance, with equipment that wasn't properly checked, filled me with an unsettling sense of failure. Either option felt like a betrayal to the fans and to the band. The weight of it hung heavy in the air, pressing down on me like a malfunction I couldn't debug.

The guilt crept in again, stronger this time. It twisted in my core, foreign and unpleasant. I didn't like it, this feeling of responsibility, of knowing that my decisions—my choice to keep things quiet and delay telling the band—had backed us into a corner. Ruby was relying on me to help her, but instead, we were on the verge of letting everyone down. The Analyst's programming had clearly downloaded something deeper, something I wasn't sure I wanted to experience. Guilt wasn't just uncomfortable; it was paralyzing. It felt like it was gnawing away at my logic, leaving me trapped in this impossible situation with no clear way out. And I hated it.

3. Xane

The morning sun streamed through the thin curtains of the tour bus, casting long streaks of light across the floor. Ruby sat across from me, her arms wrapped around her knees, eyes still red from the long night. Neither of us had slept much, and the weight of what lay ahead felt heavier than ever.

We had been running through ideas for hours—renting a replacement mixer, asking if the venue had a spare, even considering patching something together with the equipment on hand. But none of it seemed feasible. The deeper we got into this mess, the more impossible it became to ignore that sinking feeling gnawing at both of us.

"Maybe if we can just—" Ruby started, then cut herself off with a frustrated sigh. "No. That won't work either. What are we going to do?"

Her voice cracked, and I could feel her guilt tightening its grip on her. She hadn't been able to shake it since last night, and with each failed idea, the weight of it seemed to grow, suffocating any hope of a solution. I tried to stay focused, filtering through every possibility my system could generate, but the options were thin and getting thinner.

“We'll talk to the band's sound tech,” I said, trying to keep my voice calm. “Maybe he knows something we don't. He's run enough shows—he might have a workaround we haven't thought of."

Ruby frowned, shaking her head. "He'll freak out when he hears what's happening. I don't know if he'll even want to help."

I raised an eyebrow. "He likes you, doesn't he?"

She blushed, glancing away. "Yeah, I guess… We've flirted a little. But that's not going to help when we're talking about the show going up in flames."

I nodded, considering. The tech, Kevin, had been on tour with us before. He was sharp, efficient, and Ruby had his attention—whether from her charm or her flirtation, I wasn't sure. Either way, we didn't have any better options. It was time to reach out.

"Let's give him a call," I said. "We need to know what we're up against."

Ruby hesitated but pulled out her phone. I could see the nerves eating at her as she tapped his number. She pressed the phone to her ear, biting her lip as it rang.

"Hey, Kev," Ruby said, her voice soft but trying to sound casual. "How's it going?"

I couldn't hear what Kevin was saying on the other end, but Ruby's expression shifted slightly, forcing a small smile. She leaned back, glancing at me as she spoke. "Listen… we, uh, have a little situation here."

The smile faded as she explained, her voice growing quieter with every word. I watched her sink deeper into herself as the weight of the conversation pressed down. When she finished, there was a long silence. I could hear Kevin's voice faintly through the phone, serious and steady.

Ruby's face fell, and I could see her spiraling before she even spoke. "Are you sure there's no way?" she asked, her voice almost pleading.

Kevin's reply was long, and Ruby's grip on the phone tightened. She nodded as if he could see her, though her eyes were beginning to glass over. "Yeah… I get it."

She handed me the phone, her hand trembling slightly. "He wants to talk to you."

I took the phone, holding it to my ear. "Kevin, it's Xane."

"Xane," he greeted me, his voice steady but tense. "Look, I've been doing this long enough to know when a situation's gone bad. Without the mixer getting here in time a quality sound check, I'd suggest canceling the show. I know that's not what you want to hear, but it's the only way to avoid a disaster."

Canceling the show. The words hit me like a punch to the chest, but I didn't react outwardly. "Canceling isn't an option," I said. "There's got to be something else we can do."

Kevin sighed. "You're not hearing me, Xane. You don't understand what's at stake here. Without the sound check, I can't guarantee the sound will be balanced. I may make it look easy, but imagine the bass is so loud no one can hear Leila's vocals. Or worse, Lena can't hear the drums properly, and she gets out of sync with the rest of the band. We're talking about a show that could completely fall apart. I really don't want to lose my job—or my reputation—over this."

His words painted a stark picture. The kind of mistakes he was describing were the kind that could ruin a performance, disappoint thousands of fans, and destroy all the progress the band had made. I glanced at Ruby, who was watching me with wide, anxious eyes. She didn't need to hear Kevin's words to know the weight of what he was saying.

"I get it," I replied, my voice tight. "But we need a solution. We can't just pull the plug."

"I know it's tough," Kevin said, his tone softening slightly. "But if it comes down to that, you'd be better off canceling than putting on a bad show."

I glanced at Ruby again. She looked like she was on the verge of breaking down. I didn't need the aura filter to see the guilt eating away at her now. Her hands were trembling, her breathing uneven.

“We’ll figure it out,” I said, though my own uncertainty was beginning to creep in. “Thanks, Kevin.”

I hung up the phone and set it down on the small table between us. The silence that followed was suffocating. Ruby was staring at me, her face pale, her eyes filled with dread.

“He said to cancel, didn’t he?” she asked, though I could hear in her voice that she already knew the answer.

I nodded slowly. “He’s worried about how bad it could get without the sound check. He’s worried about the sound being off—bass overpowering vocals, timing falling apart. He could be under or over adjusting the entire show. He really needs a quality sound check.”

Ruby dropped her face into her hands, her shoulders shaking. “I ruined everything,” she whispered, her voice muffled by her hands. “If I had just checked the schedule, double-checked the shipment… none of this would’ve happened. And now we’re stuck. The music video really placed unexpected delays in everything. I lost focus.”

I watched her, feeling a tightness building in my chest again. Guilt. The more I was around it, the more it seeped into me, making me question every step we'd taken. Had I made things worse by stalling? Should we have told Leila right away? The idea of hiding this from the band any longer seemed like a losing battle.

"Ruby," I said softly, reaching out to place a hand on her arm. "We're not done yet. There's still time to figure something out."

Ruby shook her head, still unable to lift it, her voice strained and trembling under the weight of guilt. "We should've told them right away. Leila could've helped. Zoe… I can't keep this from her anymore." Her breathing became shallow, and she pressed a hand to her stomach, her face pale and tight with distress. "If we're going to cancel, we might even need 24 hours' notice unless one of us gets really sick... and honestly, that could be me if this doesn't get better. I feel like I'm going to throw up."

Her voice cracked, and I knew she was right. The more time passed, the more this secret weighed on both of us. Ruby's guilt was breaking her apart, and I could feel its grip tightening around me too.

"We'll tell them," I said quietly. "But first, let's see if there's anything else we can try. We owe it to them to at least keep looking for a solution."

She finally lifted her head, her eyes red and glassy. "What if there isn't one?"

I didn't have an answer for that. But I knew we couldn't give up yet. Not while there was still a shred of hope left.

4. Xane

The weight of the morning had settled heavily between Ruby and me, thick and unspoken as the minutes dragged on. Ruby had barely said a word since our conversation with Kevin, the sound technician. She had retreated into herself, looking pale and defeated, the guilt choking any energy she had left. I watched her closely as we sat in the quiet of the tour bus, her fingers nervously picking at the edge of her seat.

She suddenly stood up, her movements stiff. “I can’t take this anymore, Xane. I have to tell them.” Her voice wavered, and she glanced toward the door, her eyes darting toward the other bus where the rest of the band was. “Leila deserves to know, and Zoe’s going to find out one way or another. I need to confess before everything blows up in my face.”

I could see the conflict tearing her apart, guilt weighing her down like a physical force. She was ready to throw in the towel, and I didn’t blame her. But as she moved toward the door, the wheels in my mind clicked together, an idea sparking at the very last moment. It was ridiculous, absurd even. But it just might work.

“Wait!” I stood up quickly, reaching out to stop her. Ruby turned, confusion flashing across her face.

"There's another way. We don't have to tell them yet."

Her brows furrowed, her hand still on the door. "What are you talking about, Xane? We're out of time. There's no fixing this. Kevin even said—"

"I know what Kevin said, but hear me out," I interrupted, my voice picking up with excitement. "What if we didn't have to fix it the usual way? What if we could do the sound check without anyone realizing?"

Ruby stared at me, incredulous. "What? How would that even be possible?"

I paced back and forth, my mind racing as the idea unfolded. "Look, we need the sound check, right? The mixer won't get here until 5:00 o'clock. They can have it plugged in and set up by 6:15. The show doesn't start until 7:00. But what if we made it part of the show? A pre-show performance that gets the audience involved, while secretly letting us test everything—the mics, the instruments, the monitors. The fans wouldn't even know it was happening."

Ruby blinked, clearly not following. "A pre-show? You're going to have to explain that better."

I stopped, turning to her, a grin starting to form. "A clown skit. You and I dress up, go out on stage as clowns, and act like we're messing with the equipment in a funny, exaggerated way. Meanwhile, Kevin and the crew are doing the real sound check in the background. The audience thinks it's all part of the show, but we're actually running the technical checks we need for the show."

Ruby's eyes widened, the disbelief obvious in her expression. "You want me to go out there in clown makeup and do some comedy routine? Xane, that sounds—" She shook her head, rubbing her temples. "No, that's crazy. The band will never go for it. Leila's going to kill me if she finds out I delayed telling her for a clown act."

I could see her hesitation, but there was a flicker of curiosity in her eyes—a small glimmer of hope she wasn't quite ready to embrace. I pressed on, my excitement building. "Picture it: the audience would love it. We've never done anything like this before, so it'd be a fun way to warm them up. And while they're laughing, Kevin gets the time he needs to finish setting up the mixer. We'd be knocking out the sound check without anyone noticing, and the band wouldn't have to find out about the delay."

I could tell she was still skeptical, so I added quickly, "You don't need heavy makeup—just throw on an oversized rainbow wig that almost covers your eyes and a wild, colorful outfit. Maybe some crazy suspenders and a red clown nose. No one will even recognize you. After the skit, you'll have two or three minutes during the blackout when Zoe starts on drums and Lena comes out on keys. You can tear off the costume and be back on stage just in time to hit your bass line. It'll be a quick change, but totally doable."

Ruby still looked doubtful, but she didn't move toward the door again. She crossed her arms, biting her lip. "I don't know, Xane… What if it goes wrong? What if they figure out what we're doing? Or worse, what if the skit bombs, and we just look ridiculous?"

"It won't," I said, my voice steady, though I could sense her hesitation. "I know it sounds risky, but right now, this is the best shot we have. I'll handle most of it. You won't be out there alone. We'll plan everything—get costumes, come up with a routine that fits the band's vibe. It's not just about fixing the sound; it's about turning this into an opportunity. Think about it: a unique pre-show act that no one's expecting."

She still seemed unsure, her gaze darting toward the door. "And if it fails? We'll be a laughingstock before the show even starts."

"It won't fail," I reassured her. "Trust me. This could turn things around. We've already run out of traditional options, Ruby. This might be the only way to save the show without canceling or risking a disaster."

Ruby stared at me for a long moment, her arms still crossed. I could see the internal battle she was waging—the desire to fix this mess against the fear of making things worse. Slowly, the tension in her shoulders eased just a little, and she sighed, rubbing her forehead.

"You're really serious about this, aren't you?"

I nodded, keeping my expression calm but determined. "I am. I wouldn't suggest it if I didn't think it could work. We've got the creative freedom to pull it off, and the fans are there to enjoy themselves. They'll love it."

Ruby let out a long breath, still shaking her head, but there was a faint flicker of something in her eyes—hope, maybe. "Okay… okay, fine. I'll bite. But you better have a solid plan, because I'm not going out there looking like a fool without a real game plan."

I grinned, the weight of the morning lifting just a little. "Don't worry. I've got this. We'll plan everything down to the last detail."

Ruby raised an eyebrow, still wary. "Alright. So what exactly do you have in mind?"

I smiled, excitement bubbling up as I prepared to explain the details. This could work. It had to.

5. Xane

Tension hung thick in the air as Ruby and I walked toward the other bus where Leila, Zoe, and Lena were gathered. Ruby had agreed to give my plan a chance, but we still had to tell Leila about the mixer—and get her on board with the skit idea. As we approached the bus, Ruby shot me a glance, her eyes full of nerves.

"Are you sure about this?" she asked, her voice low.

"I'm sure," I replied, though I knew that convincing Leila wouldn't be easy. "We don't have a choice."

We stepped onto the bus, where Leila was sitting with a cup of coffee, Zoe and Lena chatting nearby. Leila looked up, her face lighting up briefly before she noticed the seriousness on Ruby's and my faces.

"What's going on?" she asked, setting her cup down. Her natural leadership instincts kicked in as she straightened up, eyes narrowing in concern.

Ruby opened her mouth to speak, but the words seemed to get stuck in her throat. She looked at me for help, and I stepped forward.

"We've run into an issue," I started, keeping my voice steady. "The mixer didn't arrive on time, which means we can't run a normal sound check before the show."

Leila blinked, her expression hardening. "What? How did this happen?"

Ruby stepped in, her voice shaky. "It was my fault. I didn't double-check the shipment, and now the mixer's delayed. Kevin already told us there's no way to do the usual sound check without it."

Leila's jaw tightened, and for a moment, I saw her frustration flicker in her eyes. She took a deep breath, clearly holding back an immediate reaction. "Okay… so what are we going to do? We can't just go out there without checking the sound. That would be unprofessional—and a disaster for the audience."

I jumped in before Ruby could spiral further. "We have a plan," I said quickly. "It's a bit unconventional, but it could work. We can run the sound check in a way that the audience won't even know it's happening."

Leila raised an eyebrow. "What do you mean?"

I explained the concept, breaking down the details piece by piece. "We'll dress up as clowns and go out on stage before the show starts. The audience will think it's a comedy skit, but really, we'll be testing the mics, the instruments, the monitors—everything. Kevin will be working the sound levels behind the scenes while we act like we're messing around. It'll look like part of the act, but we'll actually be doing the sound check in real time."

Leila stared at me, processing. I could see her weighing the idea, her natural skepticism battling with the urgency of the situation. “A clown skit,” she repeated slowly.

Ruby nodded, though she still looked unsure. “It sounds crazy, but it could work. It’s better than trying to force a regular sound check on the audience, right?”

Leila leaned back, rubbing her temples. “A clown skit… You two? You’re serious about this?”

I nodded. “Completely serious. The audience will love it, and it buys us time to fix the sound without them sitting through a boring technical check. It’s the only way to avoid a potential disaster.”

Zoe, who had been listening quietly, finally spoke up. “A skit might actually work. The crowd’s already hyped up when they arrive—they’ll be into something light and funny before the show.”

Lena chimed in, her soft voice thoughtful. “And if we can get the technical stuff sorted without them noticing, it would save us from a mess during the performance. The timing could work.”

Leila sighed, clearly not thrilled about the idea but recognizing the necessity of it. “Alright,” she said after a long pause. “Let’s hear more about how this is going to work.”

I breathed a quiet sigh of relief, thankful that she was at least open to the idea. “Here’s the plan,” I said, leaning forward. “We’ll start by playing clumsy and confused—acting like we don’t know what we’re doing with the equipment. We’ll play with the mics, test the instruments, and mess around with the sound levels, but we’ll do it in a way that’s funny. The audience will think it’s part of the act, but Kevin will be adjusting everything behind the scenes.”

Ruby added hesitantly, “We could involve the audience too—ask them silly questions, get them to make noises into the mic, pretend like we’re doing some kind of ‘audience sound check,’ but really, Kevin will be doing the actual technical check.”

Leila crossed her arms, still skeptical. “And you think the audience will go for this? We’ve never done anything like it before.”

I nodded. “Exactly. It’s new, and it’s unexpected. That’s why it’ll work. Plus, we’re buying time to ensure the sound is perfect. We’ve already missed our window for a traditional check, but this keeps things moving without disrupting the show.”

Ruby glanced at me, then back at Leila. “I’m not going to lie—I’m nervous about this. But if we don’t do something, we could end up with a terrible performance. I don’t want the fans to sit through an awful show just because we couldn’t get the sound right.”

Leila sighed again, rubbing her chin thoughtfully. “Alright. If this is our only option, we’re doing it. But I want this planned to the last detail. No improvising once we’re on stage. We can’t risk it falling apart.”

“Agreed,” I said quickly. “We’ll make sure everything’s solid before we even get close to showtime.”

Ruby and I spent the next couple of hours brainstorming specifics. We sat at the small table in the back of the bus, sketching out ideas and figuring out the flow of the skit. Ruby was still hesitant, but with every idea we came up with, I could see her getting a little more on board.

“We could start by pretending the mics don’t work,” I suggested. “You’ll pick one up and tap it, act like you can’t get it to turn on. Kevin will run the levels in real time, making it look like we’re fixing it, but he’ll actually be checking everything.”

Ruby nodded, jotting it down. “And then I could do something with the bass, like pretending the amp isn’t working right. I could accidentally crank the volume too high, and we can play that off as a gag.”

“Perfect,” I said. “We’ll add in some audience interaction too. You can call out to them, ask if anyone knows how to ‘fix’ the sound, but all the while, Kevin's getting everything sorted in the background."

Ruby was still nervous, but she was starting to warm up to the idea. We talked about the costumes, deciding on the oversized rainbow wigs, colorful outfits, and maybe even some ridiculous accessories like red clown noses.

"I'll need something easy to tear off," Ruby said, scribbling down ideas. "I don't want to get stuck in a clown suit when I need to be playing bass. Maybe some suspenders and a loose jacket I can yank off quickly."

I nodded, thinking about the timing. "You'll have a few minutes when Zoe starts on drums and Lena gets the keys going. That's your window for the quick change. Just make sure the costume is simple enough to get out of."

After we had the general flow of the skit down, we called Kevin to fill him in. To my surprise, he was more amused than skeptical.

"Wait, so you two are going out there dressed as clowns?" he chuckled over the phone. "I'm in. I've never done a sound check like this before, but I'll make sure everything runs smooth from my end."

About the Author

Mark White, the author of eight captivating novels, makes his debut in novella writing with the Xane series. A graduate of the University of Texas, specializing in Artificial Intelligence and Machine Learning, Mark blends his scientific expertise with creative storytelling to produce immersive and thought-provoking works. His latest project, ***The Xane Volumes***, is a series of 12 novellas that explore various emotions, aiming to entertain while enhancing readers' emotional intelligence. Through Xane's journey, Mark even explores the complexities of love—an emotion often debated—adding rich layers to the understanding of the human experience.

For more information and updates, visit Xane's website at Xane.me.